# CRIMES

# AGAINST

# LOGIC

# CRIMES AGAINST LOGIC

Exposing the

Bogus Arguments

of Politicians, Priests,

Journalists, and

Other Serial Offenders

## JAMIE WHYTE

### McGraw·Hill

New York   Chicago   San Francisco   Lisbon   London   Madrid   Mexico City
Milan   New Delhi   San Juan   Seoul   Singapore   Sydney   Toronto

**Library of Congress Cataloging-in-Publication Data**

Whyte, Jamie
    Crimes against logic : exposing the bogus arguments of politicians, priests, journalists, and other serial offenders / Jamie Whyte.
        p.    cm.
    Rev. ed. of: Bad thoughts.
    ISBN 0-07-144643-5
    1. Fallacies (Logic).    I. Whyte, Jamie. Bad thoughts.    II. Title

BC175.W45    2004
160—dc22                       2004053941

First published in Great Britain in 2003 by Corvo Books Ltd.

16 17 18  DOC/DOC  1 9 8 7 6 5 4

ISBN 0-07-144643-5

McGraw-Hill books are available at special quantity discounts to use as premiums and sales promotions, or for use in corporate training programs. For more information, please write to the Director of Special Sales, Professional Publishing, McGraw-Hill, Two Penn Plaza, New York, NY 10121-2298. Or contact your local bookstore.

This book is printed on acid-free paper.

For Mariam Rachel, co-gestater

# Contents

# Preface

All self-help books should begin with a confession. Here is mine: I write letters to the editor. "Outraged of London," that's me. I am getting better, though. I often don't send the letters, and sometimes I don't even write them. If I had a therapist, he would be pleased by my progress.

But I must also confess that there has been no deep reform of my character. I still want to write those letters. It's just that what gets me so riled doesn't seem to be of the least interest to the editor of the *London Times*. Nor to my increasingly fewer friends, who yawn and roll their eyes as I explain my concerns—or "rant," as the less kind among them say.

What bothers me so much?

Errors in reasoning. Fallacies. Muddled thinking. Call it what you like; you know the kind of thing I mean.

Because you have chosen to read a book with the title *Crimes Against Logic*, you may be more sympathetic than my friends and the editor of the *Times*. And sympathy is called for. The modern world is a noxious environment for those of us bothered by logical error. People may have become no worse at reasoning, but they now have so many more opportunities to show off how bad they are. If anyone cared about our suffering, talk radio and op-ed pages would be censored. Even Congress is now broadcast, as if no torment were too great.

Why are we protesters so lonely? Why don't the other consumers of all this defective thinking complain to the supplier, and to whoever else will listen, as they would if their washing machines leaked or their cars wouldn't start?

The simple answer is that most people don't notice the problem. When a car breaks down, anyone can see that it has even if he knows nothing about how cars work. Reasoning is different. Unless you know *how* reasoning can go wrong, you can't see that it has. The talking doesn't stop, no steam emerges from the ears, the eyes don't flash red. Perhaps one day someone will design a device whereby logical errors set off some such alarm, and no politician, journalist, or businessperson will be allowed to speak without having the device applied. Until that happy day, however, we must all rely on our own ability to spot errors in reasoning.

Alas, most know next to nothing about the ways reasoning can go wrong. Schools and universities pack their minds with invaluable pieces of information—about the nitrogen cycle, the causes of World War II, iambic pentameter, and trigonometry—but leave them incapable of identifying even basic errors of logic. Which makes for a nation of suckers, unable to resist the bogus reason-

ing of those who want something from them, such as votes or
money or devotion.

Many instead defend themselves with cynicism, discounting
everything said by anyone in a position of power or influence. But
cynicism is a poor defense, because it doesn't help to tell good
reasoning from bad. Believing nothing is just as silly as believing
everything. Cynicism, like gullibility, is a symptom of underde-
veloped critical faculties.

This book aims to help fill the gap left by the education system. But it is not a textbook. It is the logic equivalent of one of
those troubleshooting guides in your car or computer manual. It
is aimed at everyday users and consumers of reasoning, which is
everyone, and covers those errors in reasoning that are commonly
encountered, especially when discussing or debating controversial topics. Each of the twelve chapters is devoted to one such
fallacy.

Once pointed out, it is easy to see that they are fallacies.
Harder is spotting them in everyday life. Most of the book is
therefore devoted to discussing examples. Some are imaginary
but of such a familiar kind that you will have no trouble recall-
ing real cases you have encountered. More, however, are real
examples, drawn from politics, theology, business, and where-
ever people engage in reasoned debate—or what passes for it.

# CRIMES

# AGAINST

# LOGIC

# The Right to Your Opinion

"**K**now your rights!"

So we are advised by all sorts of well-meaners. When I was an undergraduate student, activists wanted me to know the rights that protected me against police harassment. Having dutifully learned them, I was disappointed never to encounter the expected harassment. Now I receive pamphlets telling me that I may have a right to various kinds of government assistance, including money. Alas, the result of inquiry is always the discovery that I don't qualify. As with cheap flights, conditions apply, and it seems that I am the citizen's equivalent of someone who wants to fly to Sydney at Christmas.

My poor return from knowing my rights shouldn't put you off. Knowing your rights is usually useful and we could all do better at it. How many British citizens are aware, for example, that they have a right to a good night's sleep? Well, they do.[1] In a few years, when my newborn daughter has finally got herself a decent job, I plan legal action against her.

Learning your rights can also mean discovering that you do not really have rights you think you do. This can also be useful. Suppose, for example, that you thought you had a right to do to your body whatever you like, provided you injure no one else. Such a delusion could well land you in prison convicted of drug use or assault.[2]

In this spirit, my purpose here is to stop you from believing in another right that you do not really have, namely, the right to your own opinions.

Perhaps you don't believe you have this right; then I am sorry for being presumptuous. But, you would be the first person I have met who doesn't believe it. The slogan "You are entitled to your opinion" is so often repeated that it is near impossible for the brain of a modern Westerner not to have absorbed it.

Like many other views that have at times enjoyed universal assent, however, it isn't true. You don't really have a right to your own opinions. And the idea that you do, besides being false, is forever being invoked when it would be irrelevant even if it were true.

---

1. The right was confirmed by the European Court of Human Rights in October 2001. The court upheld the claim of people living in the flight path of Heathrow airport that early-morning flights violated this right to a good night's sleep.

2. In December 1990 a group of men who, for the sake of pleasure, volunteered to have each other cut their genitals were convicted of various crimes, including actual bodily harm.

### The Irrelevant Right

Before showing that this cliché is false, let's first be clear that its common use in discussion or debate really does amount to a fallacy. It is often used preemptively, when an assertion is prefaced with the acknowledgment that "Of course, you are entitled to your opinion, but . . ." Yet its more basic use, which the above acknowledgment is intended to preempt, is defensive.

Jack has offered some opinion—that President Bush invaded Iraq to steal its oil, let's say—with which his friend Jill disagrees. Jill offers some reasons why Jack's opinion is wrong and after a few unsuccessful attempts at answering them, Jack petulantly retorts that he is entitled to his opinion.

The fallacy lies in Jack's assumption that this retort is somehow a satisfactory reply to Jill's objections, while, in fact, it is completely irrelevant. Jack and Jill disagreed about Bush's motivation for invading Iraq, and Jill gave reasons to believe that Jack was mistaken. She did not claim that he had no right to this mistaken view. By pointing out that he is entitled to his view, Jack has simply changed the subject from the original topic, the reason Iraq was invaded, to a discussion of his rights. For all it contributes to the invasion question, he may as well have pointed out that whales are warm-blooded or that in Spain it rains mainly on the plains.

As with most of our fallacies, once seen, it is obvious. Here is a simple way of putting it. If the opinions to which we are entitled might nevertheless be false, the entitlement cannot properly be invoked to settle a dispute. It adds no new information on the original matter; it does nothing to show that the opinion in question is true.

Interpreting the cliché to exclude the possibility of falsity—
that is, to mean that we are entitled to have all our opinions be
*true*—has two problems. First, it is ridiculous. Second, it does not
in fact make the entitlement to an opinion relevant in deciding
who is correct in any dispute. If Jack has a right to his true opinion then presumably Jill has a right to hers too. But then, since
Jack and Jill disagree, one of them must be suffering a rights violation; one of them has a false belief. So, even if we had the right
to true beliefs, that would only show that it is a right that is violated all the time, on precisely those occasions when our opinions are in fact false. In any dispute, to know whose right to a
true belief is being violated we would first need to work out
whose belief is false. That is, we would need to settle the *original* dispute—in the case of Jack and Jill, about President Bush's
reason for invading Iraq. And a diversion on the matter of rights
gets no one any closer to answering that question.

So, even on the strongest, and utterly incredible, interpretation of our opinion entitlement, it is irrelevant to anything else
we might be debating. Why then is insisting on one's right to an
opinion such a popular argumentative ploy?

In part, it is encouraged by an ambiguity in the word *entitlement*. It has a political or legal interpretation, by which we are
all entitled to any opinion we might have, however groundless.
But it also has an epistemic interpretation, that is, one related to,
or concerned with, truth or knowledge. You are entitled to an
opinion, in this epistemic sense, only when you have good reasons for holding it: evidence, sound arguments, and so on. Far
from being universal, this epistemic entitlement is the kind you

earn. It is like being entitled to boast, which depends on having done something worth boasting about.

So, the two senses of entitlement could not be further from each other. Yet it is too tempting to muddle them. The implied argument of the muddler runs as follows:

1. If someone is entitled to an opinion then her opinion is well-supported by evidence. (This is precisely what it means to be entitled to an opinion.)
2. I am entitled to my opinion (as is everyone in a democratic society).
3. Therefore, my opinion is well-supported by evidence.

This is a beautiful example of the fallacy of equivocation, i.e., slipping between different meanings of a word in an argument that would be valid only if the word were used with the same meaning throughout. (See the chapter "Equivocation.")

Once pointed out, it's easy to see that this confusion of the political with the epistemic notion of entitlement is a mistake. And though, strictly, that will do for the purposes of this book, I don't want to leave the matter here. Even if the cliché that we are entitled to our opinions is not employed in the truly egregious way so far discussed, it is part of a mindset that increasingly impedes the free flow of ideas and their robust assessment. Many people seem to feel that their opinions are somehow sacred, so that everyone else is obliged to handle them with great care. When confronted with counterarguments, they do not pause and wonder if they might be wrong after all. They take offense.

The culture of caution this attitude generates is a serious obstacle to those who wish to get at the truth. So it is important to strip away any bogus ideas that support the attitude, such as the idea that we all have a right to our own opinions.

## Rights and Duties

To see that there is really nothing at all to this idea that we have a right to our opinions we need only understand one basic point about rights, namely, that rights entail duties. I don't mean to endorse the fashionable slogan, "No rights without responsibilities," which is supposed to justify policies whereby the government imposes good behavior conditions on the receipt of social welfare. I mean something much more fundamental about rights: they are *defined* by the duties to which they give rise.[3]

The law gives all citizens a right to life. Your right to life means that everyone else has a duty not to kill you. This is not something that a government may or may not decide to associate with your right to life; it *is* that right. A law that did not impose on others a duty not to kill you would fail to establish your right to life. Does your right to life mean that others have a duty to feed you, to house you, or to provide you with medical care? These are hotly debated questions, but no one doubts that the answers to these questions about others' duties are what define and delimit the right to life.

---

3. For those interested in a fuller discussion of the connection between rights and duties, see P. Jones, *Rights* (Basingstoke, McMillan, 1994).

So when anyone claims a right, first ask which duties does this right impose on others; that will tell you what the right is supposed to be. And it also provides a good test for whether there is, or should be, any such right. It will often be clear that no one really has the implied duties, or that it would be preposterous to claim they should.

Mary Robinson, in her former role as United Nations High Commissioner for Human Rights, claimed that we have a human right to be healthy. Yet, without qualification it is difficult to know what she could possibly have meant. According to the World Health Organization:

> Health is a state of complete physical, mental and social well-being and not merely the absence of disease or infirmity.

Yet everyone ages and dies. And when they do, their physical, mental, and social well-being are less than complete. So the simple fact of human mortality means that everyone's right to be healthy is ultimately violated, and someone has failed to do his duty. But what could that duty be? To find a remedy for human mortality, presumably. But who could possibly bear this burden? Surely not each of us, who mostly know so little about the mechanics of human aging.

There is, of course, no unqualified human right to good health, any more than there is a human right to all those other things that it would be nice to have—such as long eyelashes and silk sheets—but which no one has a duty to provide. If she wanted to make sense of her claim, Mary Robinson should have started with the duties rather than the right. What duties does each of us have with respect to others' health or governments with respect to the health of their citizens? Then we would know what this right to good health is supposed to amount to.

## Opinion Duties

What then are the duties that the right to your opinions might entail? What am I obliged to do to respect this right? Let's start from the boldest possible demands and work down to the more humble.

Does your right to your opinion oblige me to agree with you?

No. If only because that would be impossible to square with the universality of the right to an opinion. I, too, am entitled to my own opinion which might contradict yours. Then we can't both do our duty toward each other. And think of the practical implications. Everyone would have to change his mind every time he met someone with a different opinion, changing his religion, his politics, his car, his eating habits. Foreign vacations would become as life-changing as the brochures claim.

Does your right to your opinion oblige me to listen to you?

No. I haven't the time. Many people have many opinions on many matters. You cannot walk through the West End of London without hearing some enthusiast declaring his opinions on our savior Jesus or on the Zionist conspiracy or some other topic of pressing concern. Listening to them all is practically impossible and not therefore a duty.

Does your right to your opinion oblige me to let you keep it?

This is closest to what I think most mean when they claim a right to their opinion. They do so at just that point in an argument when they would otherwise be forced to admit error and change their position. And this is also the weakest possible interpretation of the right and thus the most likely to pass the test.

Yet, it is still too strong. We have no duty to let others keep their opinions. On the contrary, we often have a duty to try to change them. Take an obvious example. You are about to cross the street with a friend. A car is coming yet your friend still takes a stride into the road. Knowing that she is not suicidal, you infer that she is of the opinion that no cars are coming. Are you obliged to let her keep this opinion?

I say no. You ought to take every reasonable measure to change her opinion, perhaps by drawing her attention to the oncoming car, saying something like, "Look out, a car is coming." By so doing, you have not violated her rights. Indeed, she will probably thank you. On matters like whether or not a car is about to crush them, everybody is interested in believing the truth; they will take the correction of their errors as a favor. The same goes for any other topic. If someone is interested in believing the truth, then she will not take the presentation of contrary evidence and argument as some kind of injury.

It's just that, on some topics, many people are not really interested in believing the truth. They might prefer it if their opinion turns out to be true—that would be the icing on the cake—but truth is not too important. Most of my friends, though subscribing to no familiar religion, claim to believe in a "superior intelligence" or "something higher than us." Yet they will also cheerfully admit the absence of even a shred of evidence. Never mind. There is no cost in error, because the claim is so vague that it has no implications for action (unlike the case of the oncoming car). They just like believing it, perhaps because it would be nice if it were true, or because it helps them get along with their religious parents, or for some other reason.

But truth really is not the point, and it is most annoying to be pressed on the matter. And to register this, to make it clear that truth is neither here nor there, they declare, "I am entitled to my opinion." Once you hear these words, you should realize that it is simple rudeness to persist with the matter. You may be interested in whether or not their opinion is true, but take the hint, they aren't.

# Motives

When my sister was fifteen, she thought she had fat thighs. Occasionally, she would demand to know, "My thighs are fat, aren't they?"

"No darling," my parents would reply, "you have nice thighs; you're a beautiful girl."

Well, that confirmed it. "You're just saying that!" was the constant refrain as my sister took our parents' protestations to the contrary to confirm all her worst fears.

My sister was committing the Motive Fallacy. She thought that by exposing our parents' motives for expressing an opinion— to make her feel better and shut her up—she had shown the opinion to be false.

But she hadn't. It is perfectly possible to have some interest in holding or expressing an opinion and for that opinion to be true. A man may stand to gain a great deal of peace and quiet from telling his wife that he loves her. But he may really love her nevertheless. It suits most to believe they are of better than average looks, and at least 44 percent of the 90 percent who believe this actually are. My sister's legs were not fat. In other words, you don't show someone's opinion false just by showing that he has a motive for holding it.

This should be obvious as soon as it is said, but in case it isn't, consider our adversarial legal system. Both lawyers in a civil case, plaintiff and defense, are hired guns. They act on behalf of whoever pays them. So it's even worse than "just saying it"; they've actually been paid to say it. Yet one says the accused did it, while the other says he didn't. So one must be speaking the truth, despite her selfish motives. To know which is speaking the truth, and hence whether or not the defendant is guilty, the jury must attend to the *evidence* presented by the lawyers, and not simply their motives for presenting that evidence. (If we followed the method of the Motive Fallacy in civil trials, they would be rather simple. Decide against the side of the lawyer who was paid more. She has the greater corrupting motive.)

Motives are relevant in deciding whether or not to believe someone only when we are dealing with *testimony*. Only, that is, when we are asked simply to take someone's word for something. Suppose your sister has let you down by marrying one of those who think it hilarious to agitate people with dramatic falsehoods. One day, he phones you and tells you your sister has won five million dollars playing Lotto. Should you believe him? No, because the probability that he is lying exceeds the probability that your sister has won Lotto, which is about one in fifteen million. If your sister had instead let you down by marrying a rigidly honest puritan, from whom the chance of deceit is much less than one in fifteen million, then you should begin to celebrate (the Lotto win, at least).

But assessing testimony is rarely the issue when motives are brought up. Normally, discussion is under way, evidence is being presented, cases are being made, and no one is being asked to take anything on someone's say so. And then, suddenly, one party starts speculating on the motives of the other.

Committing the Motive Fallacy ends a debate, not by properly refuting one of the positions, but simply by changing the subject. First, you are discussing some issue, such as whether my sister has fat thighs, and then, after the fallacy is committed, you find yourself talking about the motives of those involved in the discussion. Perhaps this is why the fallacy is so popular. It turns all discussions—be they about economic policy, religion, or thighs—into discussions about our alleged motives and inner drives. Anyone who has watched daytime television will know the delicious temptation of speculating on your own and others' psyches.

Jack informs Jerry Springer that his wife left him because she couldn't handle the fact that every woman wants him. As one of the many women who don't want Jack, Jill skips this obvious refutation and moves directly to a hypothesis as to why Jack is inclined to make such inflated boasts, declaring from the studio audience that Jack has an unusually diminutive external reproductive organ.

This is the kind of banter that entertains millions daily. It is greatly improved by focusing on the motives behind the participants' utterances rather than their truth or falsity, because the latter is usually too obvious to be interesting. But the Motive Fallacy creates an infuriating diversion when the original topic is important and the correct opinion is a matter of dispute between well-informed people—as with politics.

## Political Motives

The Motive Fallacy is so common in politics that serious policy debate is almost nonexistent. The announcement of a new policy is greeted, not with a discussion of its alleged merits, but with a flurry of speculation from journalists and political opponents regarding the politician's motives for announcing it. He wants to appease the right wing of his party, or is trying to win favor in marginal rural states, or is bowing to the racist clamoring of the gutter press, or what-have-you. If you follow politics you will be familiar with the various motives standardly attributed to politicians.

And you will be familiar with the effect. Nothing need be said about whether the policy is likely to achieve its objective, or whether that objective is sensible, or anything else about the quality of the policy itself.

Policies are treated merely as tactical moves in the game of politics. They can get you into the lead or make you vulnerable to your opponents. But their likely effects outside the game—for example, on employment, educational standards, and so on—seem to be of not the least interest to anyone playing or commentating. Occasionally, a politician declares his desperation to free himself from this horrible game and "address the issues of real concern to the people of this country." But the matter normally ends with this earnest plea. The promised issue-addressing never quite happens. And why should it? The declaration of intent will do quite nicely on its own. We want our politicians to be serious about the issues, of course. But for pity's sake, don't drag us into all the boring details!

Journalists and politicians now devote their attention to investigating the possible causes rather than the likely effects of their opponents' policies. If they can find a party donor or family friend who stands to benefit from the policy then they will have won the day. The policy is clearly rubbish.

It is, for example, a rare opponent of the 2003 American invasion of Iraq who does not enjoy speculating on President George W. Bush's motivation. He wanted to finish his father's work, steal Iraqi oil, or do the bidding of Jews who seek the removal of an enemy of Israel. It is assumed that once we are convinced of Bush's low intent, we will be convinced that the invasion was wrong.

But Bush's motives are irrelevant. He might have had the worst intentions that ever moved a leader to war. Perhaps he invaded because he was once insulted by a waiter in an Iraqi restaurant. Yet the invasion might be justified for all that. It might still liberate oppressed people and make the world safer. Good actions can be performed for bad reasons. Equally, bad actions can be well intended. Perhaps Bush really does want to liberate the Iraqi people and make the world safer. That intention does not alone make the invasion a good idea. Saddam's tyranny may now be replaced by Shiite theocratic tyranny and the world might become a more dangerous place.

It isn't only Bush's opponents who commit the Motive Fallacy. In April 2004, Senator Kerry published a "misery index" that alleged to show that life for most Americans had got worse during Bush's presidency. The index weighed factors such as net incomes, the cost of education and health care, and home ownership rates. The allegation was, however, easily answered:

> Bush campaign spokesman Steve Schmidt dismissed the index as a political stunt. "John Kerry has made a calculation that if he talks down the economy, it will benefit him politically," he said. (CBSNews.com, April 11, 2004)

Steve Schmidt was probably right that John Kerry aimed to make a statement that would help him politically. He was, after all, running for the presidency. But how can this fact possibly show that life had not really got worse for Americans during the Bush presidency?

Schmidt's response, though the kind of thing you hear all the time, is absurd. Of course those involved in a debate want to win it. That does not suffice to show that their opinions are false. If it did, entering into a debate would be self-defeating, because attempting to win would immediately show your position to be wrong. Along with John Kerry, George Bush and all other politicians would have to remain silent in defense of their opinions. Only that way could they avoid Mr. Schmidt's guaranteed refutation.

### Spotting the Motive Fallacy

The difficulty with the Motive Fallacy is not so much seeing that it is a fallacy, but spotting its instances in everyday life. It is so common that we have become desensitized, and it can be committed in subtle ways. Consider, for example, the way the news media report the publication of a "white paper" by a think tank.

The rules for doing so must be specified in some journalism manual, for they all seem to do it the same way. First, the conclusion is baldly stated: "Joining the Euro will cost three million jobs in the U.K." Then the name of the think tank: "That's according to the Foggian Society." Then the slant: "A right-leaning think tank." You wonder why think tanks always only *lean* to the left or right. (Do they have their feet nailed to the center?) Why do journalists mention this right versus left business at all? A think tank's political allegiances would be relevant only if they invited us to accept their views simply on their say so. But they don't. Their white papers are full of evidence and argument supporting their contention. To refute their view, you need to show what is wrong with the case they make.

But that would require reading the white paper, and perhaps other works cited in support of the case, and even doing a little thinking on the topic. And who has the time or energy for all that? Not journalists, who have to be experts on twenty new topics every day. Better just to point out the direction in which the think tank leans. Those who lean the same way can then agree with its finding, those who lean in some other direction can reject it.

The limits on journalists' time and their need to present information in bite-sized quantities mean that, when dealing with matters of dispute, the temptation to commit the Motive Fallacy is overwhelmingly powerful. It is done in more or less subtle ways, but attune yourself to it and you will see that the Motive Fallacy is almost universal.

Here's a tip for spotting it: watch out for the word *just*, as in "You're just saying that" or "His new education policy is just an attempt to win the student vote." Why has the "just" been included in such sentences? Everyone knows that when I say something I am saying it. What does it add to say that I am *just* saying it? Well, it is supposed to show that what I am saying is not also true, or that the education policy, besides appealing to students, is not also a good policy. The mere addition of the word *just* can, of course, achieve no such thing—it has no magical power of refutation. Nevertheless, people try it on all the time. Beware!

# Authority

"**B**ecause I say so" is something most of us were told by our parents at some time or other. Usually it was simply a threat. As an answer to "Why should I eat my peas?" for example, it is much more civilized than "Because I will beat you if you don't," and thus to be applauded. But, you may also have heard it in answer to a question about some matter of fact, such as "Why should I believe in the virgin birth?" If so, your parents erred badly, committing the Authority Fallacy.

The fallacy lies in confusing two quite different kinds of authority. There is the kind of authority your parents, football referees, and parking attendants have: the power to decide certain matters. For example, your parents have the power to decide when you will go to bed. Hence, in answer to the question "Why is 8:00 P.M. my bedtime?" the answer "Because I say so" is quite right; your parents are, quite literally, the authors of your bedtime. But it is not up to them whether or not Jesus was conceived without the help of sexual intercourse. Mary's being a virgin at the time of Jesus's birth is beyond the will of your parents, or indeed anybody else's (with the possible exception of Jesus's parents). So your father's answer "Because I say so" is quite wrong when the question is "Why should I believe in the virgin birth?" The matter exceeds the scope of his parental authority.

Yet, there is another metaphorical sense of "authority" on which the answer "Because I say so" is sometimes reasonable, even when literal authority is absent, namely, the expert kind of authority. If someone is an expert on some subject (or an authority on the topic, as it is often put) then his opinion is likely to be true—or, at least, more likely to be true than the opinion of a non-expert. So, appealing to the opinion of such an authority—i.e., an expert—in support of your view is perfectly OK. It is indirect evidence for your opinion.

We can't all be experts on everything. When laypeople sit around debating evolutionary biology, quantum physics, developmental economics, and the like, as the government's reckless education policies mean they increasingly do, one of the best pieces of evidence likely to be put forward is simply "Because Nobel laureate Joe Bloggs says so." And if Professor Bloggs himself is unfortunate enough to stumble into the wrong pub, then his saying "Because I say so" will do just as well, suffering only from an unpleasant air of arrogance.

The Authority Fallacy should now be clear. It occurs when the first literal type of authority, whereby someone has the power to make certain decisions, is confounded with the second metaphorical type, whereby someone is an expert and so likely to be right about some matter of fact.

Your father may decide when you will go to bed, what you will eat for dinner, and where you will go to school. But that literal authority does not make him an expert on human (or divine) reproduction. So, you would do well to demur when he tells you that you should believe in the virgin birth because he says so. It goes against everything you have learned in your school science classes and your father is a sales rep for Xerox, not a biologist or forensic archaeologist. Of course, he may just be threatening you, as when you asked why you should eat your peas. But, such a threat is neither here nor there. It may motivate you to believe in the virgin birth—or to say you do[1]—but it provides no evidence regarding the fact of the matter. For those interested in believing the truth the unsupported opinions of the ill-informed are of no help and are not improved upon by being offered up at gunpoint.

## The People

In a fabled past beyond memory, more than ten years ago, there was great public respect for society's authority figures. And if old introductory logic textbooks are any indication, this created a terrible problem with the Authority Fallacy. The views of parents, popes, police officers, priests, and politicians were endlessly put forward as if they carried the weight of expert opinion, despite their notorious fallibility. (The Pope's fallibility is especially notorious, on account of his tendency to deny it.)

---

1. It is impossible simply to decide to believe something, even when someone menacing tells you to. You can test this for yourself. Try to believe something you now disbelieve, say, that you are heir to the throne of Croatia or that being hit by a car will not injure you. I'll bet you can't. To believe something, you normally need some reason to think it true.

Times have changed, however, and the standing of these figures is in sharp decline. Parents are well-known by their children to be utter fools, and you will rarely hear a dispute settled with the words "Because the president says so."

So we might expect considerably less of the Authority Fallacy. Yet it thrives. New authority figures have risen to replace the old and, thankfully for books of this sort, they are just as unreliable.

The best example is The People: that is, the majority of people or, sometimes, the largest minority. The People is not merely an unreliable source that is often invoked as if it were expert. Better, it is invoked precisely because of the confusion between our two types of authority.

In a democracy, The People is the ultimate political authority. It has the power to choose the government. This may or may not be a good thing; the merits of democracy as a system for choosing the government are not our current concern. We need only note that whatever democracy's merits, it had better not be founded on the assumption that the majority of people are experts on economics, jurisprudence, international relations, and so on, because most of us are woefully ignorant on these topics. Literal authority is sometimes based on expertise—for example, most rugby referees are experts on the laws of the sport—but it isn't always. And the most obvious example of their separation is democracy. Yet the opinion of The People is endlessly invoked by politicians seeking support for their views.

Depending on which poll you go by, between 60 and 70 percent of U.K. citizens are opposed to adopting the Euro as its currency. This means that if the promised referendum were held tomorrow, the United Kingdom would not adopt the Euro. Nor should it, at least if you think the will of The People should decide this matter. But, it does not follow from this that adopting the Euro is a bad idea; The People has little idea what the consequences of adopting it would be. So it is no use appealing to this popular opposition in support of your view that adopting the Euro would harm British interests. Nevertheless, Euro-skeptical politicians do it all the time.

All democratic politicians agree that ultimate political authority lies with The People. On other matters they may disagree. One may think private schools an abomination, the other that the state should have no role in education. Each tries to convince the public that her view is right, knowing that popular opinion will decide the matter. But, "decide the matter" does not mean determine who is right. The People cannot do that; no one can by mere decision make a state monopoly on education superior to a private system, or vice versa. Public opinion decides the matter only insofar as it chooses which policy will be adopted. And the public is perfectly capable of choosing the inferior policy. If it were not, if popular opinion were invariably correct, then politicians would have no serious leadership role to play; government could be conducted by a combination of opinion pollsters and bureaucrats.[2]

The BBC's recent "Great Britons" television program was a folly based entirely on this confusion about what popular opinion can decide. A case for each of the short-listed ten greatest Britons ever to have lived was made by a celebrity advocate, and

---

2. This is precisely the direction modern politics is taking. But it is not, I suspect, because politicians genuinely believe public opinion to be infallible. Rather, it is a consequence of the professionalization of politics. Letting public opinion guide your policies, though not much increasing your chance of being right, greatly increases your chance of being elected.

then the public was asked to vote. But what was the vote to decide? Whose statue will be erected in Hyde Park? On whose birthday we will have a national holiday? Nothing but the fact of the matter. But such a fact (if indeed there are such facts) cannot be decided. It isn't up to anyone's opinion who really was the greatest Briton ever to live. And nor is the public a reliable source on such matters: Princess Diana came in third.

## Matters of Opinion

It is worth a brief digression here on matters of opinion. Some of you will think I went badly wrong above when I said that it is not a matter of opinion who was the greatest Briton ever to have lived. Surely this is a perfect example of a matter of opinion. If you are thinking this, I know what you mean. But first I must remind you of what I mean.

When I say something is not a matter of opinion I mean it quite literally; facts do not depend on opinions about them. If Princess Diana is indeed the third greatest Briton ever to have lived that is because she was pretty, kind, and so on. It is not because someone thinks she is third greatest or because the third largest group thinks she is greatest. There is nothing special about this alleged fact. No fact can be made just by being believed. So, in my literal sense, nothing is a matter of opinion.

What you (probably) mean when you say that something is a matter of opinion is that there is no objective standard by which to judge the matter, and so it is each man for his own opinion. Some think beauty irrelevant to greatness; they can choose

Churchill as their greatest Briton. Others think beauty paramount; they can choose Princess Diana. The problem is that "greatness" (as it applies to people) is a horribly unclear word, with about as many interpretations as there are people with different heroes. But clarify the word—take any one of these different interpretations of greatness—and it is not a matter of opinion who is greatest, who has the most of the quality, so understood. The debate in the "Great Britons" program was really all about what constitutes greatness, not about the various characteristics of the candidates.

Sometimes we can clarify what we mean by our vague or ambiguous words, and so explain away what was an apparent disagreement—we were just talking at cross-purposes. Sometimes we can't. Whether or not some food is good is a debate many of us have had, especially with our parents when we were children. I thought brussels sprouts were horrible but my mother insisted they were good. Neither of us could articulate any standard of goodness whereby brussels sprouts are clearly near the top, nor one on which they are obviously at the bottom. Does this make the goodness of brussels sprouts a matter of opinion?

Again, not in my literal sense. We should not conclude that I can make brussels sprouts horrible by thinking so and that my mother makes them good by thinking so, if only because that would make brussels sprouts simultaneously good and horrible—thereby violating the most basic law of logic, namely, noncontradiction. Rather, we should conclude only that I don't like them and she does, and that that is all there is to it. There is no such thing as the goodness of brussels sprouts. They have their various properties that cause a flavor sensation in my mother's mouth that she likes and one in my mouth that I don't like. No disagreement there. We get the appearance of disagreement only by projecting this reaction of ours back onto the thing that caused it: my mother saying they are good, I that they are horrible. Whether or not you like brussels sprouts is a matter of taste, but the goodness of brussels sprouts is not (literally) a matter of opinion, because there is really nothing to depend on anyone's opinion.

Of course, the colloquial use of "matter of opinion" is harmless. It merely signals the lack of a clear standard and so shows that there is probably no real disagreement, except perhaps about the meanings of words. I have gone on about it a bit only because it is important that the harmlessness of the colloquial expression should not lull you into thinking that some facts are literally matters of opinion: that something can be so just because someone thinks it is.

## Victims

Appeals to the infallible opinion of The People are the most obvious example of the Authority Fallacy at work in the modern world. As with the old authority figures, The People gains its spurious expert status in large part through fear, in this case the fear of seeming undemocratic. Disagreeing with The People is not merely bad luck for a politician who would like to be elected; it is looked upon as some kind of moral failing. It is also fear of this kind that helps our other modern authority figures transcend their hopeless ignorance.

No one wants to seem insensitive toward the victims of tragedies. When the mother of a child rape victim sobs at a press conference that the death penalty should be applied immediately to a man recently taken into custody, it takes a single-minded devotion to jurisprudence to tell her then and there about the many shortcomings of her suggested course of action. No one, however, should afford her words the weight of expert opinion simply on account of her anguish. Nevertheless, this happens all the time.

In 1995, Leah Betts, a British schoolgirl, died after taking the drug Ecstasy at a party. Ever since, newspaper articles about a suggested liberalization of drug laws also report her father's outraged reaction to the idea. Why? How has Mr. Betts's suffering made him an expert on the effects of drug laws on public health, crime, individual liberty, and so on? If it has not, why should we be interested in his opinion on the matter?

Mr. Betts is not alone in being elevated through tragedy. Victims of London's Paddington Station rail crash of 1999 are now consulted on public transport policy and there are suggestions that crime victims should be involved in sentencing the convicted. Perhaps the journalists and politicians who approach policy formation in this way are genuinely concerned for the victims whose causes they espouse, but that is beside the point. Suffering does not bestow expertise. Believing what victims believe does not make you more likely to be right.

On the contrary, the effect of suffering can be systematic error. People tend to personalize the world. Those injured in a train crash are apt to overestimate the probability of train crashes. Those who have lost a child to some disease are too ready to see its symptoms in their other children. This may be understandable in the individuals affected but it is no basis for government policy.

## Celebrities

Barry Manilow had several pop music hits in the 1970s. He probably knows the pop music business inside out. You might do well to seek his advice on how to get a record contract or how to improve a tune you have just composed.

When it comes to economics, jurisprudence, international relations, and all those other topics relevant to politics, however, Mr. Manilow is less distinguished. My little research on the topic indicates that he has no more expertise in these matters than any other randomly selected citizen. There is no apparent reason to take his advice on which presidential candidate you should vote for.

Yet during the recent Democratic primary campaign, the American public was informed, as though this was something they should bear in mind when voting, that Barry Manilow favored Richard Gephardt.

There was nothing special about Manilow and Gephardt. Other candidates also boasted celebrity endorsements. For example, John Edwards had Dennis Hopper, Howard Dean had Rob Reiner, and John Kerry had Jerry Seinfeld. Dennis Hopper, Rob Reiner, and Jerry Seinfeld are no more qualified to give political advice than Barry Manilow.

Celebrity political endorsements are the Authority Fallacy on stilts. The candidates and their campaign managers must know it; they just think the public doesn't. Let's hope they are wrong. Or, failing that, let's hope voters are so promiscuous in their celebrity worship that they get pulled in too many directions. If I were a celebrity-directed voter, for example, only Gephardt would have been ruled out. To choose between Hopper, Reiner, and Seinfeld—sorry, Edwards, Dean, and Kerry—I would have had to think about their policies.

*

Spotting the Authority Fallacy is easy. Ask yourself if the authoritative source offered up is indeed an expert on the matter. If he isn't then you should ask for the case to be made explicitly rather than merely taking his word for it. His opinion on its own is no evidence.

Be careful too that you aren't being offered *transferred expertise*. This occurs when someone who is certainly an expert in one area is offered up as an authoritative source in some wholly different area. Any opinion of Einstein's seems to rate special attention, no matter how far removed from physics. I have been told by several people that most of us use only 10 percent of our intellectual capacity. When I ask them why I should believe this, they tell me that Einstein said so. How he was in a position to know this they can't say, but everybody knows how smart Einstein was.

Well, don't be bullied. That's what I say. Einstein was indeed a very clever man, but he didn't know any more than you or I do about how much of our intellectual capacity we all use. As far as I can see, most of us are running near our limits. And if I'm wrong, then just telling me that I disagree with Einstein isn't evidence.

Being contrary isn't the path to truth, but nor is being a toady.

# Prejudice in Fancy Dress

It can be a little tricky when you have been over the matter quite thoroughly and, in the end, everyone can see that you have no good grounds for your position—or, worse, that you cannot even make it coherent. You must either give it up or else stick to it and live with everyone knowing it to be mere prejudice.

Or must you? Perhaps there is a third way. You might try claiming that it is of the very nature of what you speak that it cannot be made comprehensible to mere men. Or point out that there is much that has not been explained by the narrow rationalist approach of science; seeing the truth requires intuitive insight.

Dressed in such pieties, your prejudice now looks rather grand. Grand enough, perhaps, that no one will notice that it remains wholly unsupported by evidence.

Properly executed, the diversion will give you a reputation, not for bigotry, but for wisdom. Consider adding some literal fancy dress to your rhetorical finery. Don a simple robe or wrap a towel around your head; no one asks for evidence from people in such outfits. Sandals, facial hair, and a certain tone of voice can also be handy in the search for the high ground.

You, dear truth-seekers, would never indulge in such shenanigans. But many around you do, and their glorious aura can sometimes intimidate and bewilder the normally clear thinker. This chapter is devoted to exposing the ploys by which the prejudiced attempt to substitute sanctimony or other grand irrelevancies for evidence. I consider six. There are certainly more, but the examples considered should suffice to stimulate your intellectual immune system, so that you can identify and resist the others as well.

### Mystery

I find fish mysterious, especially their ability to breathe underwater. It has something to do with water passing through their gills, I know, but beyond that it gets rather hazy.

My finding fish mysterious tells you nothing much about fish and nothing good about me. You will conclude, rightly, that I have failed to look into the matter with any dedication. You will not conclude, I trust, that fish are intrinsically mysterious—that my ignorance is no such thing but, rather, the proper appreciation of the mystery of the fish.

Yet, on matters a little loftier than fish, this is often the moral drawn from ignorance or incoherence. Consider, for example, the orthodox Christian doctrine of the Unity of the Holy Trinity. The Father, the Son, and the Holy Ghost are three distinct entities—

as suggested by "Trinity." Yet each is God, a single entity—as suggested by "Unity." The doctrine is not that each is part of God, in the way that the FM tuner is part of your three-in-one home stereo. Each is wholly God.

And there's the problem. It takes only the most basic arithmetic to see that three things cannot be one thing. The doctrine of the Unity of the Trinity is inconsistent with the fact that three does not equal one.

It is also inconsistent with the fact that identity is a transitive relation: that if A is identical with B, and B is identical with C, then A is identical with C. If the Son is identical with God, and God is identical with the Holy Ghost, then the Son must be identical with the Holy Ghost. They are one and the same thing. But those who assert the Unity of the Trinity deny this last implication; they deny that Jesus is the Holy Ghost.

The Catholic Church—its pope, cardinals, and priests—agree that three does not equal one and that identity is a transitive relation. So they have a problem. How can the doctrine of the Unity of the Trinity be true when it is inconsistent with these obvious facts?

Well, it's a mystery. That's how. Indeed, it's a *strict* mystery. Strict mysteries are those that are of the very nature of the thing and which it is both hopeless and sinful to attempt to resolve.[1]

This response may satisfy the sheep in the congregation but it should satisfy no one with his critical faculties intact. For it sim-

---

1. Along with Papal infallibility, the notion of the strict mystery and its application to the Unity of the Trinity was settled upon at the Vatican Council of 1869–1870.

ply acknowledges the problem without solving it. The incantation "it's a mystery" does not wash away the intellectual sin of contradiction. It remains impossible both that three does not equal one and that the Trinity is a Unity. If you hold both beliefs, you contradict yourself. One belief must be wrong, and because it is necessarily true that three does not equal one, we know which it is. Cry mystery all you like; it won't stop you being wrong.

The bankruptcy of the mystery ploy is made obvious by the fact that it applies equally to any opinion you care to conjure up, however outlandish. When asked how your idea can be true given that it contradicts everything else we know and that there is no evidence for it, simply reply that it is a mystery. The Unity of the Duality, the Duality of the Quadruplcy, the Trinity of the Duality: they are all equally good candidates for mysterious acceptance, as is anything else impossible or otherwise absurd. Mystery is a completely undiscriminating license for belief. It rules out only what is coherent and well-supported by evidence, which may be why the mysterious is so fashionable with new-agers, who take belief to be a matter for unfettered self-expression.

Claiming that the Unity of the Trinity is mysterious is not only futile, it is dishonest. If you can see clearly that something is false, then there is nothing mysterious about it. The idea that the sun rises in the evening and sets in the morning is not mysterious, it's just plain false. Anyone can see this. And anyone with even the slightest education can also see that the doctrine of the Unity of the Trinity is false. You need only recognize that three can never equal one; or that if John's father is the king, John cannot simultaneously be the king. Most Christians know this much. The real mystery is why they have so little intellectual honesty.

The world abounds with genuine mystery. Most of it is quite local. The mystery of how fish breathe underwater, for example, is local to me and others of equal piscine ignorance. But many are better informed; for them, there is no mystery. Some mystery, however, is universal. What happened in the first few nanoseconds after the Big Bang, if indeed the universe started with a bang, is a mystery to everyone, including those who devote themselves to the subject. The average weight of Napoleon's hair in 1815, though a matter of little concern to most, remains a mystery and probably always will.

Some are greatly impressed by mystery. It gives them a thrilling fit of the cosmic heebie-jeebies. But all mystery, whether local or universal, whether the question is trivial or important, is a mere matter of ignorance. Nothing is intrinsically mysterious. Finding something mysterious displays no additional understanding of it, on a par with discovering that it is green or weighs two grams. It displays only a failure to understand. There is nothing noble in this failure, even if there is nothing shameful in it either. The proper reaction is to keep on studying, or perhaps to give up in defeat, but certainly not to conclude that, because the matter remains a mystery, you may believe whatever you like.

### Faith

Mystery can help the image. You must be careful what you deem mysterious. The outer reaches of science, the relationship between God and His human creatures—that's the sort of thing. You don't want to embarrass yourself by confessing to finding it a mystery how hot-air balloons stay aloft or why the tides ebb and flow with the lunar day. But keep to the right topics, and a little mystery-mongering can give off a scent of profundity heady enough to make the mind swim.

Still, you can do better. Rather than trying to obscure your prejudice, boldly declare it a virtue. You have no reason to believe what you do, no evidence, no argument. Of course not. This is a matter of faith!

Now you have captured the really high ground. Speak with a hushed and beseeching tone. Let the pain of your sincerity appear in small grimaces as you hold forth. Who but a philistine with no sense of the sacred, no respect for your deepest convictions, would expect you to provide evidence?

How scrumptious to be faithful! But utterly irrelevant to whether or not the opinion in question is true. Whatever the finer feelings associated with faith, no matter how elevated those who indulge in it, from the point of view of truth and evidence, faith is exactly the same as prejudice. Declaring an opinion to be a matter of faith provides it with no new evidential support, gives no new reason to think it true. It merely acknowledges that you have none.

When pressed, the faithful often claim that faith is required because man is incapable of knowledge in this area. This is wonderfully self-abasing: Oh God, you are so big, and I am so small, and all of that. But this self-abasement is also self-defeating. To say that knowledge is impossible is to say that, on this matter, *all* opinions must be mere prejudice. It doesn't improve things to call *your* prejudice faith.

Indeed, declarations of faith are generally self-defeating. Someone will claim this status only for those opinions he cannot defend. No one ever declares his shoe size a matter of faith, nor his mother's sex, nor the atomic weight of gold. The moment someone declares some opinion to be a matter of faith you know what to think of it.

## Odds On

The most famous argument for believing something for which you have no evidence is Pascal's Wager. Pascal claimed that it is rational to be a Christian even though the evidence available makes the position quite improbable. Because if by chance Christianity turns out to be true, then you win everlasting salvation. While if it is false—if there is no God and no heaven or hell— then you are no worse off than the correct atheist. On the other hand, if you refuse to believe, you will go to hell if you are wrong and be no better off if you are right than a Christian who is wrong. An atheist can never win, and he might lose badly. But a Christian just might win, and he can never lose badly. In other words, no matter how improbable the truth of Christianity, it's always the best bet.

There is nothing sanctimonious about this. On the contrary, it is rather tawdry. I wonder if someone who had somehow managed to make himself love Jesus on the basis of this calculation would find that love reciprocated. He certainly wouldn't if I were

Jesus. But it is a matter of little concern, since Pascal's argument is not all it is cracked up to be anyway.

Note first that what is rational to believe has here been separated from what you have any reason to think true. That is the whole point of the argument. Pascal's Wager is thus irrelevant when the question is whether or not God exists. Pascal's Wager attempts to show that Christianity is the best bet *however unlikely its truth*.

Nor, however, does the argument work on its own terms. It does not show Christianity to be the best bet. By the method of Pascal's Wager, any other doctrine that attaches everlasting bliss to agreement and everlasting agony to disagreement does equally well. The choice is not, as Pascal tacitly assumes, between Christianity and atheism alone. The choice must be made between all the different religions according to which adherents go to heaven while everyone else goes to hell, Islam, for example. Pascal provides no grounds for being a Christian rather than a Muslim. The choice between them is a 50:50 bet.

Worse, Islam is not the only heaven and hell rival to Christianity. There is also Blytonism: the view that only those who worship Enid Blyton as the creator of the cosmos will go to heaven, the rest to hell. Admittedly, I just made up this religion. But it is a *possible* religion. Why should only those religions that have so far been made up receive the benefit of Pascal's Wager? Had Pascal lived in 2000 B.C. he might have come up with his wager, and it wouldn't then have been Christianity that it defended. Nor should anyone object to the lack of evidence for Blytonism. It is the starting assumption of Pascal's Wager that the doctrine in question—Christianity, Islam, or Blytonism—lacks evidence sufficient alone to warrant belief in it.

Once you see that Pascal's Wager supports equally not only Christianity, nor even all established religions, but all possible heaven and hell religions, the game is up. For infinitely many such religions are possible. Which will you choose? Choose any one of them and the chance that you have made the best bet is not 50:50 (i.e., 1 in 2), it's one in infinity: this religion versus all the infinitely many other possible heaven and hell religions.

Each possible religion, including Christianity, is but one ticket in a lottery with infinitely many tickets. Each bet has an equal chance of being best: i.e., an infinitesimal chance. And an infinitesimal chance is no chance at all. Without some evidence, every religion is an equally hopeless bet.

## Weird Science

Jack rehearses his latest medical conjecture, claiming that diseases can be cured by the liberal application of Thames Valley mud to the chest and shoulders. Jill points out that Jack lacks the data required to support this conjecture. Neither he nor anyone else has conducted repeatable experiments with proper control groups that confirm his idea. Jack replies that the so-called scientific method has no special claim to deliver knowledge and, for good measure, throws in a few examples of questions that remain unanswered by science and a catalog of the harm it has done to humanity and the environment.

Let us grant Jack his dim view of science. In so doing, we are at least fashionable. Let's agree that scientists are sloppy, their methods unreliable, and their intentions grubby. This may please Jack, but it doesn't really help his medical conjecture. It could be that science is dreadful *and* that rubbing mud on yourself won't, in fact, cure your cancer.

If scientific methods are unreliable then many of the views we now think justified—the view that the earth orbits the sun, that light travels faster than sound, and so on—are not really justified after all. But the justification lost by these popular misconceptions is not thereby transferred to the unsupported conjectures of those who don't like science. Jack may be quite right about the bankruptcy of science; his hypothesis about mud curing cancer remains mere conjecture.

Again, this is obvious from the fact that the antiscience ploy can be used in defense of any opinion at all, including contrary opinions. Jack claims that human civilization was kick-started by space aliens who taught our ancestors about fire, the wheel, and so on. Jill claims we were taught these things by a breed of talking donkey that has since become extinct. Jack and Jill can both rail against the tyranny of science, but they can't both be right about the history of human civilization.

Not everyone who enjoys dabbling in conjecture wants to appear antiscience. For them there is always quantum physics. No one can doubt that quantum physics is scientific; people have even received Nobel prizes for it. But look! It's completely crazy. So, you see, crazy ideas like mine are perfectly scientific.

It is a rare foray into gobbledygook that does not begin with a tribute to quantum physics. For example, Lyall Watson begins his *Supernature* by claiming that:

Science no longer holds any absolute truths. Even the discipline of physics, whose laws once went unchallenged, has had to submit to the indignity of the Uncertainty Principle. In this climate of disbelief, we have begun to doubt even fundamental propositions, and the old distinction between natural and supernatural has become meaningless.[2]

What the old distinction between the natural and the supernatural was we are not informed, but, because it has become meaningless in the new climate of disbelief, my guess is that it was the distinction between what we have reason to believe and what we have no reason to believe. And once that distinction has become meaningless, well . . . fasten your seatbelts kids.

Before considering the real implications of quantum physics for scientific reasoning, it is worth giving *Supernature*'s opening paragraph a little attention, because it is a beautiful example of a style popular with those who like their ideas fast and loose.

Begin as you mean to continue. Start with an outrageous and obvious falsehood: "Science no longer holds any absolute truths." How about this one? Light travels faster than sound. This is a scientific discovery, and it is true. It is pointless to add that it is absolutely true, because truth is always absolute. Unlike bald men, of whom some can be balder than others, true statements cannot differ in their degree of truth. A statement is either absolutely true or not true at all. Consider our example truth. How could this be partially true? Light travels either faster, slower, or

---

2. L. Watson, *Supernature* (London, Hodder and Stoughton, 1973), p. ix.

at the same speed as sound. In none of these circumstances is it partially true to claim that light travels faster. It is either absolutely true or it is false.

"Absolutely" adds a nice touch of confusion to the falsehood, but the confusion doesn't stop there. By saying that science "no longer" contains absolute truths Dr. Watson implies that it once did. How can this be? What is absolutely true surely can't become false. The mind boggles. Have the laws of nature changed since the days when science contained absolute truths?

What Dr. Watson presumably means is that some of what scientists once thought true they now think false. It is a shame he doesn't simply say this, but I suppose the banality that scientists revise their opinions isn't the kind of observation with which one starts a book on the supernatural.

One obvious falsehood and two serious confusions, and we are only on the first sentence. Now for sentence two: "Even the discipline of physics, whose laws once went unchallenged, has had to submit to the indignity of the Uncertainty Principle." The claim that the laws of physics enjoyed a period of unchallenged acceptance will entertain most historians of science, but that is a matter of little concern compared to Dr. Watson's misuse of the Uncertainty Principle. This is not a principle, as strongly suggested by Dr. Watson's paragraph, that states that the laws of physics are uncertain. It is, rather, the principle that you cannot simultaneously measure certain properties of a subatomic particle, its position and its velocity, for example. At any time, one of these facts about a subatomic particle must be unknown.

How this principle can possibly show that science contains no truths or make the distinction between the natural and the supernatural meaningless is an utter mystery to me. If those who introduced the principle had given it a different name—the Measure Exclusion Principle, say—then perhaps Dr. Watson would not have tried on this preposterous nonsequitur. It's a pretty cheap play on the word *uncertain*. But for those who swallow this pill, the rest of *Supernature* will no doubt be passed quite comfortably.

Dr. Watson's flirtation with the Uncertainty Principle attempts to draw on a vague but widespread idea that, by being strange but true, quantum physics has shown that normal standards of coherence and observational evidence are obsolete in science. We have already noted that the obsolescence of a standard confers no support for opinions that do not meet it. But it is still worth noting that quantum physics has not really had any impact at all on the standards of scientific inquiry.

The Uncertainty Principle, which so excites gobbledygookers, is part of the Copenhagen Interpretation of quantum physics. The Uncertainty Principle and other elements in this interpretation of quantum physics run contrary to some of our common ideas, at least about the world of medium-sized objects that we can observe with our unaided sense organs. Superposition—the thesis that quantum objects are simultaneously in states we would normally take to exclude each other—and the role of observation in bringing about a so-called collapse into just one of these states are the hardest to understand. By most people's lights, the Copenhagen Interpretation of quantum physics does indeed look weird.

But this should give no comfort to those who willfully embrace weird ideas. The strange elements of the Copenhagen Interpretation are the kind of things that physicists, if they accept them at all, do so only under duress. And the duress is observational evidence. The Copenhagen Interpretation is not willfully peculiar. It is an attempt to explain the observational evidence.

The weirdness of quantum physics is not an example of an intellectual free-for-all but just more of the tyranny of the scientific method. More important, though, it may contain what we find hard to understand; the true interpretation of quantum physics does not involve paradoxes, that is, statements or sets of statements that contradict themselves. It cannot, because paradoxes are impossible, no matter how small the things we consider. The philosophy of quantum physics is concerned in large part with showing that its paradoxes are merely apparent.

Internal consistency and observational evidence are required no less by quantum physicists than by any other scientists. It is only because we continue to care about consistency and evidence that there is any problem of interpreting quantum physics, or, indeed, any intellectual problems at all.

Anyone who thinks that her favorite weird ideas—about reincarnation, astral travel, or whatever—are intellectual bedfellows with quantum physics ought to read some of the latter. She will find the experience disillusioning.[3]

## But Still

Many middle-class racists adore members of the race they despise. At least that's what they say. "I have many black friends.

---

3. Many readers may find it fascinating. The topic is extremely difficult, and properly to engage with it requires a good understanding of statistics. For readers with determination, I recommend Michael Redhead, *Incompleteness, Nonlocality and Realism: A Prolegomenon to the Philosophy of Quantum Physics* (Oxford, Clarendon Press, 1987).

Indeed, I like most of the blacks I've met. But you must admit that, on the whole, they're lazy and violent." So familiar is this line of thought that the last part may now be omitted. Just say you have many black friends and everyone will take the lazy and violent bit as understood.

This approach to the question of racial characteristics is only reasonable. The reasonable person recognizes the strength of the case for the contrary view. You don't just blunder in, pointing out that every black you've ever met is a fiend. To conclude on the basis of such experience that they are an evil bunch would display the worst kind of bigotry. You can never personally know enough nasty blacks properly to draw the conclusion that they are generally nasty. No. You must first admit that most of your limited experience suggests blacks to be decent and likeable. Only then is it reasonable to conclude that they are, on the whole, irredeemable villains.

Though the ploy is most famously employed by racists, it is enjoyed by bigots of all sorts. Attend closely to the evidence for the contrary view, nod appreciatively while it is rehearsed, perhaps even throw in some evidence of your own, and then utter the magic words: *But still.*

Yes, your boss offered you extra training, asked if you needed time off, gave you several warnings. But still, he was always out to have you fired.

*But still* is a tool of logical inversion, the mirror image of *therefore.* Where the evidence would incline anyone to say "Therefore, grass is green," you may instead say "But still, grass is red." With *but still* in your logical arsenal you need fear no evidence. If it fits your hypothesis, you have *therefore.* If not, *but still.*

And, as usual with our prejudice ploys, it's *but still* anything. The evidence suggests that grass is green. Fine. But still, grass is red, or blue, or any color you like.

Colloquial English provides alternatives to *but still* for those keen on ignoring evidence. You have just heard formidable evidence suggesting your opinion is wrong. Alas, you have *but still*-ed twice already this morning. Try "yes, yes, of course, but at the end of the day, when all is said and done, you must admit that [insert opinion here]."

Besides straight reversal, watch out for exclusion. The technique is parodied in Monty Python's *The Life of Brian*. Reg, leader of the People's Front of Judea, asks his comrades "What have the Romans ever done for us?" expecting a great chorus of "Nothing!" Instead, one of the revolutionary brothers mentions the flat Roman roads. "Yeah well, there is the roads," Reg acknowledges. "But apart from that, what have the Romans ever done for us?" "The sanitation," pipes up another, and others mention yet more Roman improvements to their lives. But Reg never gives up: "All right, but apart from the sanitation, medicine, education, wine, public order, roads, fresh water system, and public health, what have the Romans ever done for us?"

You see it outside movies most often when businesspeople and politicians want to exclude what would naturally lead to conclusions they don't care for, such as that they are no good at their jobs. "Setting aside an exceptional $200 million loss on an investment in pork-belly futures, last year's profit was a healthy $100 million."

But that $200 million *was* lost on pork-belly futures, so that the annual result was a $100 million loss, not a $100 million profit. Why should we shareholders not count it? It was indeed exceptional. It was exceptionally bad! That is no reason to exclude it from an assessment of management performance. Would exceptionally good performance be excluded on the same grounds? "If you ignore the $200 million profit on llama farming in California, our result was a disappointing $100 million loss," is not the sort of thing you hear as often.

The evidence can be mixed. Some of it points in one direction, some in another. In such circumstances, it can still be rational to draw one conclusion rather than another: on balance, the evidence favors this one. But it is cheating to tip the balance by ignoring evidence that doesn't suit. Evidence should be ignored only when it is unreliable. A scientist gets a positive result but then discovers that the equipment in his laboratory is faulty. So he discards the result. That's fine. And utterly different from acknowledging something and then ignoring it for no reason except that you wish to draw a different conclusion.

## 'Tis Evident

Some things are said only when false. The popular sign, "Authentic Olde English Pub," is one of them. Authentic old English pubs do not display such signs. "It goes without saying" is almost as bad. What goes without saying goes without saying. If you feel the need to mention that something goes without saying it probably doesn't.

You should be similarly suspicious when someone tells you that his opinion is self-evident or obvious. If it is obvious, why would he feel the need to point out that it is? Just say it. Its obviousness will do its own work. And if it is not really obvious, then his claiming it is probably means he is trying to obscure the fact that he has no evidence at all—like those poor men who, unable to think of anything witty to write in their self-portrait for the personal ad, simply inform us that they have a good sense of humor (GSOH). Evidence, like a GSOH, is always more convincing displayed than merely claimed.

The boldest example of a bogus claim to obviousness is the second paragraph of the American Declaration of Independence:

> We hold these truths to be self-evident, that all men are created equal, that they are endowed by their Creator with certain inalienable Rights, that among these are Life, Liberty and the pursuit of Happiness.

Perhaps these statements are true, perhaps not. We can't get into that here. But they are not self-evidently true.

Take the claim that all men are created equal. *Equal* must here be used with some special meaning. For it is not normally grammatical to say that things (other than numerical values) are simply equal. They must be equally something: equally tall, equally green, equally obtuse, or something. What does it mean to say simply that people are equal? But if a statement is hard even to understand it surely cannot be self-evident.

Entitled to equal treatment under the law, or something along those lines, is probably what was meant. Yet principles of jurisprudence of this kind are debated even today. This one must have looked pretty bizarre to most reading it in 1776. Then, many Americans, including the authors of the Declaration, owned slaves. Can it really have been self-evident that, in a country full of slaves, everyone is entitled to equal treatment under the law?

Candidates for self-evident truth are statements such as, "I've just fallen in a puddle" or "This tea is hot." Grand principles of justice that few have ever heard before and can hardly understand aren't really the sort of thing. It is, indeed, pretty hard to come up with any kind of evidence or support for them at all—as philosophers working in the field will tell you. Hence, the temptation to declare them self-evident.

*

The ploys considered in this chapter have a common failing that may help you to recognize their relatives. They apply not only to the opinion for which they are adduced, but to any opinion at all including its direct contradiction. And that makes them worthless. Something that supports every opinion equally thereby supports none at all.

Another test for the hopeless absence of evidence is what might be called *moral positioning*. Does the opinion's defender seem a little precious on the topic? Perhaps it hasn't yet come to a *fatwa*, but she may in more subtle ways suggest that those who wish to keep friends in polite society ought to back off. Hurt feelings are in the cards if the matter is pushed too far.

Such sentiments are rarely roused in someone who can defend his position with sound argument and evidence. Tell someone his feet don't look like a size nine and he will gladly prove you wrong by displaying an old shoe box or setting his feet against someone's whose you accept are a nine. It is only when someone cannot defend his opinion, and is not interested in believing the truth, that she will attempt to stifle discussion with good manners. Those who take religion, politics, and sex seriously do not adhere to the general prohibition on discussing these topics. And they don't take offense when they are shown to be wrong.

If you start to feel during a discussion that you are not so much incorrect as insensitive, then you are probably dealing with a respectable bigot.

Only a thug would expose him.

# Shut Up!

Suppose you suggest to your husband that subsidized llama farming would set California's economic woes aright, and he takes issue with you on the ground that you are an old bag. He has clearly committed a bad mistake in logic (and perhaps others as well). Because, even if you are an old bag, it may still be true that subsidized llamas are exactly what California needs. The two statements are not inconsistent. Your husband has not won the argument; he has not refuted your opinion.

In another sense, however, he probably has won the argument. Unless you have a really exceptional devotion to agrarian reform in California, his remark will have diverted you from your original hypothesis and on to other topics. Indeed, if you are of normal sensibilities, you may well have been left altogether silent by the remark. The brute has won the argument, in the sense of getting you to drop your point, getting you to shut up.

Of course, you gentle readers would never marry such a thug, nor keep his company even as a casual acquaintance. But you will, nevertheless, have encountered remarks that serve only to shut you up, without showing that your position is wrong. And precisely because your friends and family, and your political representatives too, are nice, well-educated people, these remarks will be less obviously brutal and so easier to confuse with a proper refutation. In this chapter, I discuss three kinds of abuse that are commonly passed off as refutation. They are slightly better etiquette than calling your wife an old bag, but no better logic.

## Shut Up—You're Not Allowed to Speak

The Turner Prize is the United Kingdom's most prestigious contemporary art award. Each year, the candidates are whittled down to a short list of four artists whose work is then displayed in a London gallery before the winner is announced a few weeks later. The event always stimulates a kind of ritualized debate on the merits of contemporary art. Some say it isn't art at all; others say it is art but extremely bad, while its defenders say it is good art precisely because it elicits the first two responses. This ritual is enormous fun for its participants and so not normally allowed to politicians. In 2002, however, the notoriously indiscreet new Minister for Culture, Kim Howell, dipped a toe in. He scribbled on the gallery comments board that the short-listed works were unworthy of the prize. Or, more precisely, he claimed they were "cold, mechanical, conceptual bullshit."

This could not stand. What might the paying public (the tax payers) think? Keith Tyson, the eventual winner of the Turner Prize, came to the rescue, rejecting this criticism by pointing out that the minister couldn't do better. Or, more precisely, that a drawing by Mr. Howell, sold for charity, was "laborious . . . insipid, a piece of middle-class kitsch so lacking in life that it could win the *Daily Mail*'s competition for 'real' art."[1]

Harsh, but irrelevant. Mr. Howell did not claim that *he* ought to win the Turner Prize. He claimed that the short-listed artists ought not. His artistic abilities are neither here nor there. It could perfectly well be true that he is a dreadful artist *and* that the short-listed works are "cold, mechanical, conceptual bullshit." Exposing Mr. Howell's artistic limitations may have won the argument, in the sense of getting him to keep quiet—it must, after all, be painful to be sneered at by artists when you are Minister for Culture. But this childish "you couldn't do better" response is still only abuse and intimidation, not refutation.

So is the equally common "you can hardly talk" reply. Fat Jack tells Jill that she has put on some weight lately and she replies that he can hardly talk. Well, perhaps Jack *can* hardly talk; perhaps he is so obese that it takes all his strength to move his lips. But Jill may in fact have put on weight for all that. Jack's weight is irrelevant. She hasn't refuted him, just abused him.

Nothing could make the poverty of the "you can hardly talk" reply more obvious than its old-fashioned equivalent: "That's the

1. *The Guardian*, February 4, 2003.

pot calling the kettle black." The kettle is black! If the pot is black too, well, so much the worse for the pot, but that hardly changes the kettle's color.

These are clichés, but don't think that the underlying mistake is made only by those who use them. In public debate, the idea that you can refute a view by claiming its advocate is not entitled to speak is pervasive, especially on race-related issues.

Fat Jack may not comment on others' obesity on account of his own. When it comes to race, however, the matter is reversed. Racial characteristics, or issues affecting a race, should be commented upon only by members of the race concerned. As I write, an important case concerning the constitutional status of preferential admission for black university applicants (*Grutter v. Bollinger*) is being heard by the U.S. Supreme Court. I have just read an article by the Associated Press that canvasses the opinions of President Bush's three top-ranking black secretaries and advisers (Colin Powell, Rod Paige, and Condoleeza Rice): one is for preferential admissions, one is against it, and the other is unsure. *Why* they hold these positions the article does not say. I gather also, from television and newspapers, that both sides of the debate are keen to have their case publicly supported by blacks. There is, you will admit, nothing unusual about this. That is how it is nowadays with the public discussion of sensitive issues.

There may be a good reason for this, and for all the other complex but reasonably well-understood speech entitlements in our modern society. But, whatever the reason, it cannot be that only those entitled to speak can speak the truth. Whether or not a preferential admission policy is constitutional has nothing to do with the race of those who say it is or those who say it isn't, as the fact that both sides have black advocates makes clear.

Most of us delight in making critical observations about the characteristics of our compatriots or the foibles of family members. The same observation made by an outsider can be offensive. But if it was true in your own mouth, then it is also true in the outsider's. We mustn't confuse being sensitive with being right. Nor rudeness with error.

## Shut Up—You're Boring

An opinion's entertainment value suffers from wear and tear. When you hear something over and over it is likely to become dull. But its truth value does not. An opinion that was true on its first outing does not become false through overuse. Yet it is a common objection to an opinion—as if it constituted a refutation—that we have heard it all before. Opinions and arguments are dismissed as pedestrian, plodding, obvious, tired, and so on. Such objections might be to the point if we were discussing radio plays or striptease shows, but they are irrelevant when considering the truth of an opinion. The speaker may be shamed into silence, but his opinion is not thereby shown false.

On the contrary, most truths are apt to become familiar and unexciting. No one thrills to the idea that the earth orbits the sun like they used to. But this new blasé attitude has not altered the structure of the solar system. Equally, much fiction is surprising and not in the least dull to read, but it remains fiction for all that.

The best refutations also tend to draw on facts that are tediously obvious. How better can you refute an opinion or theory than by showing it to be inconsistent with something well-known to be true? Of course, being inconsistent with *anything* true, well-known or not, refutes a theory. But we avoid unnecessary squabbling when we choose a theory-contradicting truth not itself in dispute.

For example, Marxism, in its early prediction-making phase, made a number of claims about the inevitable effects of capitalism. One was that the workers' revolution would first occur in the society most advanced along the capitalist road, namely, England. Another was that the workers would become alienated from the products of their own labor. I'm not certain what this latter claim means, but I'm sure it is an interesting idea that people could debate all day long. This makes it a bad route to the refutation of Marxism. Perhaps the workers are not really alienated from the products of their labor and so Marxism is wrong. But who can say? Better to stick with the perhaps tedious but pretty well-known fact that the workers revolution first happened in feudal Russia, not capitalist England.

It is a strength, not a weakness, of an attempted refutation that it draws on the mundanely familiar. Yet, in the academic humanities (literary studies, sociology, and the like), where being sexy is the fashion, refutations are often dismissed on precisely this ground. For example, most humanities students and many academics claim that truth is culturally relative, so that what is true depends on what is the generally held view in the culture concerned. This relativism about truth is inconsistent with some very well-known facts, such as the fact that the earth orbited the sun in A.D. 900. Cultural relativism entails, on the contrary, that in A.D. 900 the sun orbited the earth. This is what people then believed, so it was then true. Contradicting something this well-known has always struck me as a serious problem for relativism, and I have pointed it out in many debates. But, trust me, it doesn't bother the advocates of relativism in the slightest. It is, after all, such a predictable objection, drawing on such a banal observation. Relativism, on the other hand, is excruciatingly sexy—all the more so for flying in the face of well-known facts. Far from flinching, most relativists reply that yes, in A.D. 900 the sun did indeed orbit the earth. What fun!

Perhaps the most perverse variation on this "you're boring" theme is the "you would say that" reply. It's perverse because it attempts to hold it against someone that he is consistent in his opinions. Libertarian Jill offers up her latest idea for a free-market solution to a social problem—reducing unemployment by removing the minimum wage, let's say. Socialist Jack replies that this would exacerbate already unacceptable levels of income inequality in the workforce. And Jill answers, with a roll of the eyes, "You would say that."

Yes, Jack would say that because he is a Socialist. This is the kind of view Socialists hold. Jack is simply sticking to his political ideology when he worries about income inequality. The question is whether or not the Socialist view that Jack is so admirably consistent in advocating is correct. And Jill's observation clearly fails to answer this question.

Of course, if they are old friends, Jack and Jill will have had the Libertarianism versus Socialism debate a million times, and Jill's "you would say that" may just be a way of suggesting that they not have it again. No one is going to put Jill's proposal into effect, because they are simply chatting in the pub. They can afford to move on to their vacation plans without getting to the right answer about the minimum wage.

When politicians chat about such issues, however, things are (or should be) more serious, because the country is run according to the outcomes of their policy debates. Politicians are not friends who might as well sidestep, with a glib remark, what would otherwise be a tedious debate. They lack Jill's excuse for the "you would say that" reply. But watch out for it in politics and you will see it often: "This is the senator's familiar old carping," "Have we not heard enough from Representative Watson on the plight of dairy farmers?," and so on. The consistent advocacy of an idea is turned on its head, so that instead of counting as an intellectual virtue in the advocate, it counts as a defect in the idea advocated. The idea must be wrong, look how boring its proponents are.

Politicians should try to remember that, even after the introduction of television cameras into Congress, entertainment is not their primary function. The truth can be boring. It is a little disappointing perhaps, like the fact that your husband watches too much football on television. But, if you love it you will put up with the dull moments.

### Shut Up—You Sound Like Hitler

Mass murder is something of a lottery. Lenin hasn't done so badly. I recently had a drink in the popular Lenin Bar in Auckland, New Zealand, decorated with red stars and black and white images of the great Communist. Very fetching. Hitler bars, on the other hand, seem to be in short supply. Lenin is also doing all right in the world of ideas. Communism isn't what it was among intellectuals, but you cannot yet dismiss a political or economic view simply by pointing out that it was held by Lenin. Hitler, on the other hand, is like a reverse Einstein. If you can associate someone's opinion with Hitler, or the Nazis more generally, then goodbye to that idea.

"That's just what Hitler thought!" would, on its own, constitute a successful refutation of an opinion only if everything Hitler thought was false, which it clearly wasn't. Even the worst among us has many true beliefs.

Everybody knows this, of course, and so rarely will someone try to refute, for example, the view that Berlin is in Germany on the ground that Hitler thought so. The bogus refutation is reserved for those opinions that can more plausibly be associated with Hitler's wrongdoing. Anyone who advocates using recent advances in genetic engineering to avoid congenital defects in humans will pretty soon be accused of adopting Nazi ideas. Never mind the fact that the Nazi goals (such as racial purity) and genetic engineering techniques (such as genocide) were quite different from those now suggested. The association is good enough to do the trick. No one wants to look like a Nazi.

Even if the accusation doesn't completely silence the advocate of genetic engineering, it will certainly put her on the defensive and elicit the now standard preamble about the many (unspecified) ethical problems raised by genetic engineering, how carefully it must be regulated, and so on and on until all this effort to dissociate herself from Nazism has had exactly the opposite effect in the mind of the listener. Apologizing is such a guilty thing to do.

Hitler is a great favorite, but *refutation by association* does not require his services. Other objectionable individuals or groups will do just as well if the association can be made to stick. Religious fundamentalists are quite good. Victorians aren't bad either. I have heard certain strands of feminism dismissed on the ground that they are a kind of neo-Victorian puritanism, which nicely gets in both of the above associations. Whether the feminist ideas in question are correct is irrelevant, but it should do to shut up feminists, who like to think they are modern and secular.

As usual, the mistake is more common than explicit statements of it. It is what underlies a widespread gang membership approach to belief. Many hold political or religious opinions not because they have any reason to think them true but just because they like the associations. There is, for example, a kind of package deal of beliefs for lefty students. They are anti–free trade, pro–environmental protection, in favor of more redistribution of wealth, a bit feministy (but not as much as they used to be), they think animals have rights, and, though they are not adherents of any established religion, they do have a kind of religious sentiment that inclines them to a bit of waffling on the topic. Why do they hold all these positions? Simply because that is the current lefty package—and aren't all intelligent, concerned young people at least a bit lefty?

Lefty students aren't the only ones who pick their politics and religion like fashion accessories or gang insignia. Most social groups, even those that are not explicitly ideological, have membership opinions. The ideas of a Beverly Hills country club member can probably be predicted with more accuracy than those of a Democrat.

We live in a mobile society. There is no need to feel bound to one style of thinking forever—the ideological equivalent of those sad sixty-year-old men with their yellow teeth, wingtip shoes, and Elvis hairdos. As our lefty students age, so will the ideas that suit them. Just as they will move to the suburbs, take up golf, and don pressed khakis, their political ideology will drift rightward.

And thank goodness! Left-wing rhetoric sounds so naïve coming from a forty year old. Do act your age (and shut up).

# Empty Words

Academic seminars can be pretty dry. Before committing yourself to two hours of doodling and moving your weight from one buttock to the other, it is best to get some idea of what you're in for, hence the common practice of posting brief descriptions of seminar papers before they are given. Here is an example:

> In this paper, dealing principally with *The Winter's Tale*, Harald Fawkner explores phenomenological descriptions of a recognisable hiatus between affectivity and sensibility. He proposes that, in moving from the great tragedies to the so-called romances, Shakespearean drama becomes increasingly preoccupied with an understanding of emotion as quasi-aesthetically disconnected from

the apparently conditioned mundanity of cause and
effect, representivity and determining contextuality.[1]

Something has gone wrong with the language in this synopsis.
The sentences are strictly grammatical and all the words are from
the English language, or at least a small extension of it, but it is
nearly incomprehensible. You can tell that Professor Fawkner
is going to talk about Shakespeare's plays, but what he has to
say about them remains a mystery. What, for example, can
be expected from someone who "explores phenomenological
descriptions of a recognisable hiatus"? Should I wear a raincoat
to the seminar? How does someone who thinks "emotion is
quasi-aesthetically disconnected" from something—be it the
apparently conditioned mundanity of cause and effect or anything
else—differ in his opinion from someone who believes this dis-
connection to be genuinely aesthetic?

Or perhaps nothing has gone wrong with the language here.
Perhaps the point of such language is not to communicate your
ideas to the reader but simply to give an impression of being
learned while saying almost nothing at all. Perhaps all this
inscrutable verbosity is meant to shroud the banality of the ideas.

Who knows about Professor Fawkner? Perhaps his paper,
unlike its synopsis, was a model of clarity, full of big ideas about
the Bard. I will remain forever in the dark on the matter because
I now make it a policy never to attend academic literature seminars. Most of you readers will also attend few, even if not by pol-

---

1. I discovered this example—which is by no means unusual in certain academic circles—in the "Pseuds Corner" section of *Private Eye*, April 4–11, 2003.

icy. Yet you will not be saved from the many popular mutilations of language that serve to obscure the message or the fact that there isn't one. Such language abounds in business, politics, and academia—wherever people have an interest in sounding as though they are cleverer and more cram-packed with insight and good ideas than really they are.

As with the threat of terrorist atrocities, a vigilant population is required to stamp out obscurantism. This chapter is intended to assist the struggle by providing a guide to obscurantist language. The examples discussed do not cover the full range—that would require a book in itself—but they will allow you to identify the most threatening cases. Whether on the tube, at a political party conference, or in a boardroom meeting, you will know when it is time to alert friends and colleagues and evacuate the intellectual area.

## Jargon

Jargon can aid clarity. When it does, it is usually less pejoratively called *terminology*. For example, the special terms of economics—net present value, price elasticity, and so forth—are indispensable to the subject. Because, unlike their everyday relatives —worth, spendthrift, and so forth—they can be measured with some precision. By stipulating precise meanings for such terms, and especially by making them measurable quantities, economic ideas can more clearly be expressed and tested.

Economics is not special in this respect. The role of such terminology in articulating and testing theories is a hallmark of rigorous science. This may explain why those who wish to pass off their ideas as enjoying scientific rigor are inclined to employ as much jargon as possible—management consultants, for example.

Management consultants sell advice for large sums of money. The advice can be useful, which, hopefully, explains the fees. But, there is a certain insecurity in the young men and women who prepare the presentations of management consultancy firms. The advice, however good, often seems too simple. Your company has an over-capacity problem. Then you had better increase sales volumes or cut capacity. How might you increase sales volumes? Lower your prices, perhaps, or open new sales outlets, or something else. How likely to work are these options? New sales outlets would be too expensive because customers for your type of product are very price-sensitive. Then lowering prices is the best bet.

That's the sort of idea involved. It's fine but . . . it just doesn't sound clever enough from Ivy-League and business-school graduates charging $5,000 a day. It needs sexing up with some jargon.

Jargon in management consulting involves the substitution of bizarre, large, and opaque words for ordinary, small, and well-understood words. The substitution is no more than that. Consultese brings with it no extra rigor, no measurement precision lacking in the ordinary language it replaces. Where terminology in the sciences aids clarity and testability, consulting jargon shrouds quite plain statements in chaotic verbiage.

Though few of you readers will be management consultants, many will have been exposed to their language, because it now permeates the business world and, increasingly, journalism and politics. Certain words must be included wherever grammar permits, and sometimes even where it doesn't. Fashions change, but "leverage," "intellectual capital," "benchmark," and "moving forward" are perennial favorites. Suppose, for example, that the way your company is run means that the good ideas of the staff are not heard by the managers, and that other companies do better in this respect. This becomes:

> Benchmarked against best-in-class peers, intellectual capital leverage reveals significant upward potential moving forward.

Some readers will need help with the translation. To "benchmark against" means to compare with. If you compare your height with the three men closest to you in the crowd and find that you are tallest, you say, "benchmarked against greatest geographic proximity persons, I am a top quartile height performer."

"Best-in-class peers" are good examples of something similar to the object of analysis. For example, if you are Dutch and standing in a crowd in Barcelona, then it may be premature to declare yourself a top quartile height performer. You may stand a head above all these Spaniards but a head shorter than a top quartile Dutch height performer. That is, you may be a runt when benchmarked against best-in-class peers.

"Intellectual capital" is stuff you know that can be useful for the business. For example, the ability to add and knowledge of tax law are part of an accountancy firm's intellectual capital. This is a peculiar use of the word *capital*, because the things a company owns, which it may or may not use for its operations, are normally called its assets. Capital is what's left after you deduct a company's liabilities from its assets. So, "intellectual assets" would seem more apt. But there's no accounting for pointless jargon, I suppose.

"Leverage" must wait until last, because it leaves the strongest taste.

"Reveals upward potential" means "could be improved," provided we want more of it, as we do in this case. "Reveals upward potential" is better than "could be improved" because when you get more of something, the number that measures how much you have is moving upward. "Upward potential" implies numerically precise measurement, whereas "could be improved" does not, and so is always to be preferred, even when such precise measurement is not possible, as in this case.

The modifier "significant" here means "a lot." Something revealing significant upward potential could be improved a lot. But "a lot" is not the kind of thing said by people who receive significant sums of money for their advice.

"Moving forward" means "in the future." It is gratuitous in any sentence with a future tense verb, such as, "You will receive our bill, moving forward." And it is also gratuitous in our example sentence, because this talks of *potential*, which is always realized in the future. Nevertheless, "moving forward" is a useful addition because more words are generally helpful, even when they add no information, and because it puts a positive slant on things. We are not, you will note, moving backward.

Now, let's return to the masterpiece of consultancy jargon: "leverage." Nothing better embodies consulting's parody of the proper use of terminology in the sciences. Normally, you get leverage when you use a lever to apply force to some object. The lever rests on a pivot and the degree of leverage is measured as a ratio—the length of the lever on one side of the pivot to the length on the other. This notion of leverage has a perfectly natural extension in finance, often called gearing. If you invest

$50,000 of your own money and borrow the rest to buy a $500,000 home, then your leverage (or gearing) is 9:1—your debt is nine times your equity. The extension is sensible because the financial case, like the physical case, involves a precise ratio.

In consultese, however, "to leverage" just means "to use." "Leverage" is leveraged in sentences that have nothing to do with ratios, like this one and like our example sentence. Management consultants even talk about leveraging each other. If I am the boss and find myself doing too much work, then the consultants in my team are not leveraging me properly. (I'm not making this up.) In short, consultants take an item of proper terminology, with a precise measurement-based meaning, and use it as a synonym for a word with a more general, everyday meaning. They make the precise vague, thereby reversing the proper use of terminology.

Why do this? Well, "leverage" sounds impressively technical and, like "moving forward," it puts a positive slant on the sentence. When you leverage things, you get out more than you put in. Don't just use it, leverage it!

It's pathetic but nonetheless popular. And, you can sympathize. Who would feel comfortable charging $250,000 for:

> Companies like yours make better use of their employees' knowledge.

Spotting jargon is difficult. The first sign is that you can't understand what is said. But, that could be because perfectly respectable terminology is being used and you don't know what it means. It could be your ignorance not their bullshit. That is the fear, lurking at the back of most businesspeople's minds, that keeps them nodding appreciatively as all this babble cascades out over the charts and tables of their advisers' presentations. No one wants to be exposed. We are all up to speed on the latest business science!

It takes a solid command of the subject—be it business, political policy, or literary studies—to spot the difference between terminology and jargon, to tell when the unusual words add nothing to precision but are merely bamboozling synonyms for ordinary language. That is why insiders are best positioned to blow the whistle. Alas, they are also most likely to have an interest in the silence of the whistle, which may explain why it is so rarely heard.

## Weasel Words

Making a clear claim is a dangerous business. Say something plainly and you can just as plainly be shown wrong. If Jack claims that the price of gold will rise next week, then we will all know he was wrong when it falls. If Jack wants to avoid exposure, he will couch his prediction in the language of the financial advice profession, saying something like:

> If U.S. interest rates stay below the 3 percent benchmark and market sentiment remains positive, then gold could rally to new highs in the near term.

The conditional nature of the statement is helpful. When the price drops, who will remember exactly what the conditions were? Whatever they were, they obviously weren't met. In this case, they are certain to have been unmet if the price falls. For they include market sentiment remaining positive, and the only test for market sentiment being positive is that the price goes up.

So Jack's prediction is simply that if the price of gold goes up, it will go up. Which is at least true. Such conditions, though rendering them worthless tautologies, are nearly always present in the predictions of financial advisers.

"In the near term" is good, too, because it means the time for your prediction to come true is never quite up. Even if the price falls next week, it might rise the week after next, and that is still the near term, isn't it?

But best of all is "could." If you substitute "could" for "will," or "could be" for "is," then you can never be wrong. After all, anything is possible. When the price of gold goes down, then Jack will still be right: it could have gone up.

"Believed to be" is also a useful substitute for "is." Jack may declare that gold is believed to be a good investment without fear of refutation by its price falling. He didn't say who believed it and he certainly didn't say the belief was true.

The cost of making statements immune from error by the inclusion of these weasel words—"could be," "believed to be," and the like—is that they are thereby rendered empty. Insofar as we take these words seriously, the statement no longer tells us anything informative. Of course gold *could* go up in price. We want to know if it will. I know some people *believe* gold is a good investment. Everything, no matter how stupid, is believed by someone. The question is whether or not this belief about the price of gold is true.

Users of weasel words hope that they will be overlooked until things go wrong. They are a kind of insurance policy. If gold goes up, good: Jack said it would. If it doesn't, well never mind: he only said it could.

Not only predictions are hedged with weasel words; all sorts of alleged findings are couched in this language. Here is a typical example from a *Daily Telegraph* article apparently reporting recent findings on the health effects of cannabis (May 2, 2003). The article is titled, "Cannabis 'could kill 30,000 a year.'" In it, Professor John Henry is quoted:

> Even if the number of deaths is a fraction of the 30,000 we believe possible, cannabis-smoking would still be described as a major health hazard.

First, note the expression "the 30,000 [deaths] we believe possible." Professor Henry cannot strictly mean what he says here. Thirty thousand deaths was well-known to be possible before any research was conducted. Any number is *possible*. It would be newsworthy to discover that 30,000 is possible only if there had been some prior reason to suspect that it is impossible. But what could have made it impossible?

By saying that he believes 30,000 cannabis-caused deaths to be possible, Professor Henry is probably trying to say that he believes 30,000 to be the actual number, but that he isn't certain. His evidence, though not conclusive, makes 30,000 the best estimate of the number.

There are standard statistical measures of how much confidence we should have in a hypothesis in the light of experimental findings. Familiarity with these statistical measures cannot be assumed in laypeople and so they are not normally reported in newspaper articles. But, throwing in the word *possible* to indicate that the proper degree of confidence in the hypothesis is less than one (i.e., not certainty) is worthless. No hypothesis is made certain by experimental results. "Possible" cannot distinguish those hypotheses enjoying strong evidential support from those with only weak support.

Worse, if the available evidence makes 30,000 the best estimate of the number of cannabis-caused deaths—a number worth publishing—then it would be more idiomatic to say that this is the "likely" or "probable" number. Then again, likely is so much riskier than possible. If, on further investigation, 5,000 turns out to be the real number, then 30,000, though still possible, won't seem to have been very likely. Why give hostages to fortune by speaking plainly?

Or perhaps Professor Henry is speaking plainly. Perhaps his research does not warrant more confidence in the 30,000 number than is expressed by saying that it is merely possible. Then you wonder what all the fuss is about. Why should anyone be interested to read about this nonfinding in the newspaper?[2]

But we need not worry too much about the number of deaths caused by smoking cannabis. It is allegedly irrelevant to Professor Henry's conclusion that cannabis smoking is a major public

---

2. Professor Henry introduces no new evidence regarding the health effects of cannabis. (See his editorial in the *British Medical Journal*, Vol. 326, May 2003, pp. 942–943.) Rather, he arrives at his 30,000 figure by assuming that cannabis and tobacco smoking increase the probability of death equally. Then he applies the annual tobacco-smoking death rate of 0.9 percent to the 3.5 million cannabis-smoking population to arrive at his 30,000 figure. His reasoning has two problems. First, the figure of 3.5 million cannabis users is based on poor data, as you would expect where a criminal activity is involved. Second, no account is taken of the different smoking habits of tobacco and cannabis users. Cannabis smokers typically smoke less each day and maintain the habit for a shorter period of their lives. Professor Henry was wise to say that 30,000 is only a possible number of deaths caused by cannabis.

health problem in the United Kingdom. "Even if the number of deaths is a fraction of 30,000 cannabis [is] still a major health hazard."

This is a strange opinion. Suppose the fraction is ¹⁄₁₀₀. Suppose, that is, that the annual death rate is only a hundredth of the 0.9 percent Professor Henry assumes, that is, only 0.009 percent. Then the chance that you will die this year from smoking cannabis is the same as the chance that the U.S. government will default on its financial obligations.[3] Hardly a major health hazard. By failing to limit the fractions he has in mind, Professor Henry makes his assertion obviously false.

But not if you remember the weasel words! Professor Henry did not actually say that no matter what the number of deaths caused by cannabis, it is a major health hazard. He said that no matter what the number of deaths caused by cannabis, it would "still be described" as a major health hazard. And on this he is probably right. I would not describe a 0.009 percent chance of death as a major health hazard, but that is irrelevant because Professor Henry doesn't say who would do the describing. Since anyone can do it, including Professor Henry himself, he seems guaranteed to be strictly right in what he says.

He is right, however, at the cost of having said nothing informative. What looked like some interesting new finding on the mor-

---

3. The credit rating of the U.S. government is AAA. Credit ratings give a rough measure of the probability that the rated institution will default on its financial obligations in a given time period (usually one year). AAA rated institutions have a 0.01 percent (1 in 10,000) chance of defaulting. This is close enough to risk free to be so-called in the financial world: the rate of interest received on loans to the U.S. government is called the risk-free rate.

tality effects of smoking cannabis turns out to be no more than Professor Henry informing us that, no matter how many deaths may or may not be caused by cannabis smoking, he will insist on describing it as a major health hazard. Which is a curious and interesting fact, but about Professor Henry rather than cannabis.

## Hooray Words

Are you in favor of justice? I'll bet you are.

Do you think it just that those who earn more should be compelled by the government to give a portion of their income to those who earn less? I dare not guess. Many think this redistribution of income essential to a just society; others think it is simple theft.

Everybody favors justice. They disagree only about what is just and what unjust. *Justice* is in this sense a hooray word. Declare that you are in favor of it and everyone will cheer his agreement, even when he disagrees with you on every particular question of what is just.

Besides justice, there is peace, democracy, equality, and a host of other ideals that everyone embraces, whatever they believe these ideals to consist in.

And then there are the boo words: *murder*, *cruelty*, *selfishness*, and so forth. Everyone agrees that murder is wrong, no matter how much they disagree about which killings are murder. Many of those who earn the disapproval of legislators by beating their children also disapprove of cruelty; they think it cruel to raise children without giving them the discipline of corporal punishment.

If you wish to make your ideas clear, you will avoid using hooray or boo words without first saying how you understand them. For example, it would be worthless for a politician to claim simply that, when it comes to the issue of wealth redistribution, she favors a just policy. We all do. What we want to know is which policy she thinks just.

Though worthless, this is precisely the approach favored by many politicians. Using unexplained hooray words is an easy way of winning the agreement of the electorate. Say you seek a fairer society. Everyone listening will have his idea of where society is currently unfair and, if he likes your straight white teeth or open-neck shirt, he will be inclined to think that you must be addressing his pet peeve. You can win at least the sentiment of agreement without having to say anything that might be held against you later. Or without even having to know what you mean yourself. A politician may have no specific concept of justice or equality or of anything at all. He has only the hooray words and seeks subsequent guidance on the details from focus groups, public opinion polls, and a process of policy trial and error if elected.

Tony Blair announced New Labour's education policy at the 1996 party conference by declaring that his "three priorities in government would be education, education, and education." Hooray! Education not once but three times! The children are saved!

Everyone agrees that education is important. People disagree about which policies will lead to the highest standards being achieved, consistent with resources being properly allocated to other matters, such as defense, justice, and transportation. Education may well be Mr. Blair's three priorities but what does he propose to do about it? That is the question any voter should want answered. "Education, education, education" just blows sunshine up our arses.

A simple test for substance in political statements is whether anyone sane would disagree. If a politician declares it her aim to make the people of Britain healthy, wealthy, and wise, she tells you nothing useful. How will you use this information to choose between her and her opponents, who almost certainly seek the same things? In a healthy democracy, where voters demand the information required to make sensible choices between parties and candidates, political discussion would focus on the difficult and controversial issues where reasonable people disagree. The commitment to justice, peace, and all the rest would, literally, go without saying.

## Quotation Marks

Suppose I claim that the minister's refutation of the allegations was feeble. Then I have said something confusing. To refute a claim is to show that it is wrong. If the minister has indeed shown the allegations to be wrong there is nothing feeble about that. I should say that the minister's "refutation" was feeble. The quotation marks around "refutation" make it clear that I do not intend to claim that it was a real refutation. He did not actually show the allegations wrong; it was merely an alleged refutation. If I were speaking instead of writing, I might register my skepticism by saying "refutation" with a sneering tone or adopt the now popular gesture of wagging the first two fingers of each hand beside my ears like a demented rabbit.

There is usually nothing confusing about this use of quotation marks. It is an economical device for saying that the so-and-so referred to by the word in quotation marks is only an alleged or a so-called so-and-so. But some writers apply the device so inconsistently or so excessively that it becomes impossible to know what they mean.

Here is an example of inconsistency from Dr. Miriam Stoppard's bestselling *New Babycare Book*. Chapter 2 includes a guide to the reflexes you should expect from a healthy newborn baby. One is the crawling reflex, whereby the baby is apt to pull her knees toward her chest when placed on her belly. In the section's subheading this is called the "crawling" reflex. The quotation marks, I assume, are intended to indicate that the baby is not really crawling—the movement merely resembles crawling. But, in the text, Dr. Stoppard states that this "reflex" will disappear as soon as the baby's legs uncurl and she lies flat. Which leaves one wondering whether the crawling reflex is not really crawling or not really a reflex, or not really either—which would make it a peculiar name for the behavior concerned.

In Dr. Stoppard's case the inconsistency is a harmless slip. Quotation marks play a small role in the *New Babycare Book*. But other authors use of quotation marks to express skepticism— sneer quotes—is uncontrolled and bamboozling. It is especially common in writers who adopt a critical attitude toward others' alleged intellectual successes. Sociologists of science, for example, are often reluctant to say that a scientist discovered something or solved some intellectual problem. They prefer to say only that the something was "discovered" or the problem "solved." The philosopher and historian of science, Imre Lakatos, provides a nice example:

> Using a false theory as an interpretative theory, one may get—without committing any "experimental mistake"— contradictory factual propositions, inconsistent with experimental results. Michelson, who stuck to the ether to the bitter end, was primarily frustrated by the inconsistency of the "facts" he arrived at by his ultra-precise measurements. His 1887 experiment "showed" that there was no ether wind on the earth's surface. But aberration "showed" that there was. Moreover, his own 1925 experiment also "proved" that there was one.[4]

The difficulty created by this incessant use of sneer quotes is that the reader becomes unable to tell not what the author does not mean by the word within quotation marks, but what he does mean.

---

4. I. Lakatos, "Falsification and the Methodology of Scientific Research Programmes" in Lakatos and Musgrave (eds.), *Criticism and the Growth of Knowledge* (Cambridge University Press, 1970), p.164, note 12. I owe the reference to David Stove's *Popper and After: Four Irrationalists of Science* (New York, Pergamon Press, 1982), Part I of which provides a hilarious exposé of the linguistic obscurantism of four prominent philosophers of science.

When Lakatos places "showed" and "proved" in quotation marks you assume that he means to imply that the experiments did not really show what they were alleged to. Why not? Perhaps because we now know that the things allegedly shown are not true. You can't prove what isn't true, you can only "prove" it. But this cannot explain Lakatos's use of sneer quotes, because he uses them both for experiments that "showed" that there is no ether wind and for experiments that "showed" that there is. Or, perhaps the experiments only "showed" these things because they were not conducted correctly. This, however, is ruled out explicitly by Lakatos, who says that Michelson's measurements were ultra-precise and that these issues arise even when no experimental mistake is committed. (Although, again he confuses us by placing "experimental mistake" in sneer quotes.)

You begin to get the feeling, after reading a few pages of Lakatos, not so much that he does not care for these particular claims to proof, but that he does not like any. Words that imply some kind of intellectual achievement or failure always get put into sneer quotes.

Why then does he persist in using such words? The reader knows what he doesn't mean by "proved" or "showed"—namely, proved or showed—but what on earth *does* he mean by them? I defy anyone to say, after reading the quoted passage, what Lakatos thinks was the logical relation between Michelson's ether theory and his experimental results.

Lakatos's mainly 1970s works were in the avant-garde of sneer quotes abuse. By the 1980s, it had become standard practice among intellectuals keen to distance themselves from the cultural imperialism of enlightenment rationalism. Nowadays, no decent person would use words implying anything evaluative— *improved, normal, mistake,* and so on—without hiding their shame under some nice modest quotation marks. They use these words of course; it is almost impossible to discuss anything without doing so. But they don't really mean them. Or, at least, not the nasty bits.

# Inconsistency

In the days before I left New Zealand for England, my father's dinner conversation took on a new theme. Every night he would invite me to enjoy the meal, saying, "You won't get food like this in England. All they ever eat in England is . . ." One night it would be roast beef that the English only ever ate, the next night steak and kidney pie, then meat loaf, and so on: every night a new dish that was the only one available in England.

After a few days my mother became exasperated with this line of conversation and decided to move on from eye-rolling to a proper refutation. On this night my father had claimed that all they ever eat in England is fish and chips.

"What about Bisto gravy-mix?" was my mother's devastating intervention.

"What about it?" came back the fearless head of the house.

"Bisto gravy-mix comes from England," she informed him.

"Yes, but you don't put gravy on fish and chips!"

It is not clear to me, when I reflect on this dinner banter, whether senile dementia was beginning to get the better of my father or whether he was just having more fun infuriating us. In any event, you will admit that my mother had got him. The statement that they eat Bisto gravy in England is inconsistent with the claim that all they eat in England is fish and chips. It is impossible for both statements to be true. And because it is true that the English eat Bisto gravy, it must be false that all they eat is fish and chips. My father's reply, that even the English don't put gravy on fish and chips, though made triumphantly, simply confirmed the refutation.

My father's response to the Bisto gravy-mix crisis was willful inconsistency. He could hardly deny the fact that Bisto is English. But he refused to be put off his generalization. He simply maintained an inconsistent position: that all they eat in England is fish and chips and that they eat Bisto gravy in England. His position was hopeless, but if you've lived through the depression, World War II, and a 66 percent marginal tax rate, what's a little self-contradiction to be frightened of?

Such blatant inconsistency is unusual. When inconsistent statements have been made explicitly, most feel the need to reject at least one of them. Only the infuriating, the mad, or the intoxicated will straightforwardly assert what everyone can see to be an impossible position: one that, since it is internally inconsistent, must contain a falsehood.[1]

---

1. It is worth noting the difference between two ways of being inconsistent. Statements can be contrary, so that at least one of them is false and possibly both. Or they can be contradictory, so that one must be false and the other true. "Jack was born in Jamaica" and "Jack was born in Barbados" are thus contrary, while "Jack was born in Jamaica" and "Jack was not born in Jamaica" are contradictory.

But when inconsistency is not so obvious, it is common. Consider, for example, the statements:

Evil exists,
and
An all-good and all-powerful god exists.

Their inconsistency is not immediately obvious, but inconsistent they are. An all-good god would want to avoid any evil he could and an all-powerful god would be able to avoid any evil he wanted. Hence, if there were an all-good and all-powerful god, there would be no evil. So the existence of such a god is inconsistent with the fact that there is evil. (By "evil" I mean nothing metaphysical: toothache, trailer park–destroying tornadoes, or torture will do.)

Nevertheless, most Christians, Jews, and Muslims believe in both. They have either failed to draw out the implications of their belief in an all-good and all-powerful god, or they have been convinced by one of the many bogus theological attempts to show this belief consistent with the existence of evil.[2]

In this respect, there is nothing special about the religious. We all hold inconsistent beliefs because we have failed to draw out all the implications of some of our beliefs and so failed to see that they are inconsistent with others we hold.

Recognizing all the logical consequences of everything we believe is beyond the intellectual capacity, and the will, of us

2. There have been many attempts to show that the existence of evil is consistent with the existence of an all-good and all-powerful god. For a survey, please see J. L. Mackie, *The Miracle of Theism* (Oxford, Clarendon Press, 1982).

humans. An extraordinary number of mathematical truths can be derived from just a few basic axioms easily understood by us all. Yet most of us are incapable of making the derivation. And even those mathematicians up to this task will surely fail elsewhere: in the various nooks and crannies of their minds will lurk beliefs inconsistent with one another.

To demand perfect consistency would be futile. But the impossibility of perfection doesn't mean we should have no expectations whatsoever. Most of the inconsistency that pollutes popular debate would take the merest effort to identify and eliminate. For example, the inconsistency in the common pair of beliefs—the government should cut taxes and the government should spend more—isn't very hard to identify.[3] It is surely not too much to ask of a voter.

Especially once the inconsistency has been noticed. This chapter aims to help by identifying two contexts in which inconsistency often goes unnoticed. There is more inconsistency about than fits into these categories, but the rest follows no pattern I can discern.

---

3. These statements are not directly inconsistent, since the government has two sources of funds other than tax: profits from state-owned businesses and borrowing. However, few recommend a return to the days of widespread state ownership of business, because such businesses usually contribute losses rather than profits. And borrowing to fund spending, though advisable during a recession, is not sustainable over the long run, at least if the rate of borrowing exceeds the rate of economic growth. In the long run, spending must be funded by tax revenues. It is this view, held by most voters, combined with the beliefs that the government should cut taxation and that it should increase spending that makes an inconsistent threesome.

## Implied Generalizations

Scientists and bold conversationalists state their generalizations explicitly. They come out and plainly say things like "Any object acted on by no force will remain stationary" or "All they eat in England is fish and chips." This explicitness makes it easy to tell what would be inconsistent with the generalization: an object moving though acted on by no force or an English food that isn't fish and chips, like Bisto gravy.

Such explicitness, however, is rare in public or private debate. More often generalizations are only implied. And then it is harder to see when something is inconsistent with them. The inconsistency, like the generalization, is only implicit. Before considering examples of implicit inconsistency, we should first see how generalizations can be implied.

Suppose that, when asked why he thinks homosexuality should be illegal, Jack replies, "Because it is denounced in the Bible." Jack's reply implicitly invokes a generalization: namely, that everything denounced in the Bible should be illegal. Without this generalization, it would not follow from homosexuality's biblical denunciation that it should be illegal. If Jack is consistent, he will seek the criminalization of everything denounced in the Bible, including adultery, banking, wearing mixed-fiber fabrics, and eating anything from the sea that lacks fins or scales, such as lobster.[4]

---

4. These biblical prohibitions are to be found, respectively, in *Leviticus* 18:20, *Deuteronomy* 23:19, *Deuteronomy* 22:11, and *Leviticus* 11:9.

This is how generalizations most often occur in discussion. They are the implied underpinning of inferences, from "condemned in the Bible" to "should be illegal," from "human" to "mortal," from "man" to "cheating bastard." Those who have made the inference can usually see and, if they are honest, admit that their inference is undone if they concede a counter-example to the implied generalization: properly legitimate behavior condemned in the Bible, an immortal human, or a faithful man.

Usually, but not always.

In 1985, Jack's argument for the criminalization of homosexuality was often heard in New Zealand. A private member's bill to decriminalize homosexuality was before parliament and a great national debate on the topic was raging in town halls, newspapers, and on talk radio. At one of these town hall meetings, I asked an evangelical Christian leader of the campaign to keep homosexuality criminal, who had just employed Jack's argument, if he thought adultery should be illegal because it is also condemned in the Bible. "No," he replied, "that would be taking things too far." His moderation won rapt applause from the assembled faithful.

Yet, this moderate response rendered his objection to legal homosexuality groundless. If it is not generally true that what the Bible condemns should be illegal—if other men's wives, 3 percent interest, lycra-cotton shorts, and lobster thermidor may all be enjoyed by law-abiding citizens—then why not homosexuality too? Far from successfully defending his objection to legalizing homosexuality, the evangelist had defeated himself. He admitted a counter-example to the generalization on which his argument depended.

You might think that contradicting yourself in public would make for successful rhetoric only when you are an evangelical preacher addressing your flock or someone enjoying similarly uncritical attention. Alas, no. Tony Blair is the acknowledged master of modern political rhetoric, and, if the appearance of moderation calls for it, he will not hesitate to contradict himself in public. His position on fox-hunting provides an example.

In September 1999, the Prime Minister sought to reassure nervous blood-sports enthusiasts that he was not some kind of fanatic. He wrote an article in the *Daily Telegraph* insisting that, though he opposed fox-hunting and would vote for its criminalization, he was a stout defender of shooting and fishing:

> There will be no ban on the country pursuits of shooting and fishing. Let me make this perfectly clear. As long as I am Prime Minister, I guarantee that this Government will not allow any ban. We will not do it.

Mr. Blair claimed in his article that fears of a legislative threat to shooting and fishing had been whipped up by members of the pro-fox-hunting lobby wishing to exaggerate the threat to the rural way of life. Perhaps. But the only slander against Mr. Blair required by their scaremongering was the accusation of consistency.

Which principle makes fox-hunting an abomination but does not equally condemn shooting and fishing? Most anti-fox-hunting activists think that it should be banned because it is cruel. Yet are not fishing and shooting also cruel? Is a hook through the head or a bullet in the gut so much kinder than a hound at the throat? What is the difference between fox-hunting and shooting that means that the former is impermissible but the latter to be defended at all costs?

By supporting shooting and fishing, Mr. Blair tacitly agrees that cruelty to animals is not a sufficient ground for criminalization. But then, what *is* his objection to fox-hunting?

Fox-hunting certainly differs from fishing and shooting in some respects. For example, it is practiced primarily by wealthy non-Labour voters on horseback and is hated by left-wing members of the Labour party who often feel Mr. Blair does too little for them. But these don't seem relevant to the question of whether or not something should be illegal. Unless he articulates his position more clearly, giving a principle of jurisprudence that shows why fox-hunting should be illegal but not fishing or shooting, then it will appear that Mr. Blair's position on blood sports, besides being moderate, is arbitrary and inconsistent.

No, some will say, it is simply pragmatic. Pragmatism is a rightly vaunted quality in politicians but it does not require inconsistency. Suppose that Mr. Blair is not as inconsistent as it appears and really seeks a ban on all blood sports. He knows, however, that he cannot now get a ban on fishing and shooting, because they are too popular. Yet, he might get a ban on fox-hunting, providing he promises to protect fishing and shooting. This, then, is what he should do, because some progress is made —foxes will be saved from the indignity of being killed by hounds (now they will be shot)—while none is made by insisting on an unachievable ban on all blood sports.

Nowhere in this pragmatism is there any inconsistency. The position on blood sports is consistent and the legislative compromise perfectly sensible under the circumstances. Inconsistency arises only when Mr. Blair claims, not that he is willing to protect shooting and fishing as part of a bargain, but that he believes there is no ground for their criminalization. It is the attempt to conceal the pragmatism, if that is what was going on, that gives rise to the inconsistency. Of course, it may be pragmatic to conceal your pragmatism; that is a matter on which I must defer to experts, such as Mr. Blair. But if so, the price of pragmatic pragmatism is deception and incoherence, which seems too high.

## Weird Ideas

Despite the decline of organized religion, weird ideas remain popular in the West. What do I mean by "weird," you may ask. But even as you ask you will know the kind of thing I have in mind. All that new-age stuff: reincarnation, astrology, numerology, homeopathy, and the like. It's all weird.

Weird but true, the defenders of some slice of this fruitcake will tell you: they have evidence. Consider reincarnation. The evidence offered up is certain people's knowledge of the past that, it is argued, could only be acquired by memory of experiences in a previous life. During a session of hypnosis aimed at putting her in touch with her former self, Jill remembers that Julius Caesar had a heart-shaped mole on his left buttock. Then a quick check of the historical sources and, lo! The conqueror of Egypt did indeed possess the alleged beauty spot. Yet Jill had never studied the matter, never seen the historical sources in question. Honest. So, you see, she *is* the reincarnation of Cleopatra.

The defenders of reincarnation are frustrated by the way this kind of evidence is dismissed by the scientific establishment. They are so wedded to their dogma, to their so-called scientific method, that they cannot see the truth before them.

In fact, rejecting the conclusion that reincarnation explains Jill's uncanny knowledge of Caesar's beauty spot is a simple matter of consistency. Reincarnation is inconsistent with much of what most scientists currently believe about the mind. It assumes, for example, that the mind (or, at least, the memory) survives bodily death, goes through a period of disembodied existence, and later reinhabits a new body. But we have good reason to believe that the mind depends on the functioning of the brain, so that it cannot survive bodily death. To accept reincarnation is to reject the view that the mind depends on the brain.

Of course, this idea *could* be mistaken. But the evidence for it is formidable. Anecdotes about what people claim to remember under hypnosis, on the contrary, make poor evidence. They can be explained in many ways that do not require us to abandon our well-supported theories about the relationship between the mind and the brain. It is always more likely that Jill is lying about not having prior knowledge of Julius Caesar's beauty spot, or that the hypnotist suggested it to her, or even that it was a lucky guess, than that modern neuropsychology is wrong and Jill really is the reincarnation of Cleopatra.

Weird ideas are not intrinsically weird. They are weird because they are inconsistent with established views of the laws of nature. When you accept them, you reject the established view, or at least part of it. This is not something to be done lightly, because the established view has a mountain of evidence to support it. Those who reject it, by believing in reincarnation, astral travel, or astrology, take a bold step. For the step to be rational, the evidence for their favored theory must be stronger than the evidence for the established view that it contradicts. Yet, it is rarely more than a collection of anecdotes about people with unlikely knowledge or who recover from a flu in double-quick time.

Besides being intellectually frivolous, advocates of weird ideas also often contradict themselves—by continuing to believe in the laws of nature that their weird idea contradicts. Consider homeopathy.

This is the idea that a disease can be treated by administering doses of a substance that, in a healthy person, would cause the disease. Because large doses of these substances cause unwanted side effects, the dose is made minute by a process of repeated dilution. Alas, the dilution of homeopathic medicines is so great that the resulting liquid is simply water, with not a trace of the originally added substance.

Homeopathologists acknowledge this fact but insist that the prior addition of the active agent and process of dilution give it curative powers lacked by water without this history. They thereby contradict a principle that I have never heard anyone seriously call into question, and which I am sure that even a homeopath would not directly deny, namely, that objects with the same properties have the same causal powers, regardless of how they came to have those properties. For example, if Jack and Jill both weigh 140 pounds, then by standing on properly functioning scales each will cause it to register "140 pounds." It makes no difference if Jack has recently gained weight and Jill lost weight. The same goes for anything else, including samples of water. It makes no difference what used to be in the water. If both samples are now purely water they will have the same effects on the health of those who drink them.

If you believe that homeopathic medicines have effects lacked by water, then you put yourself in an unenviable position. You must either deny that homeopathic medicines are just water, despite the fact that their dilution makes the presence of the active agent impossible. Or, you must deny that objects with the same properties have the same effects. Unless you deny one of these plausible ideas, you contradict yourself by believing in the efficacy of homeopathic medicines.

## Real Contradictions

Most are distressed to find that they have contradicted themselves. But not all. Some will declare that yes, their position is inconsistent, but is not the world itself full of contradiction?

That all depends on what you mean by "contradiction." I hate to say that because the meaning of "contradiction" is quite plain. So, there ought to be no question about what you mean by it or of the world being contradictory.

Statements are contradictory when the truth of one entails the falsity of the other, when, if one is true, the other must be false. "Jack is fat" and "Jack is not fat" are thus contradictory. That is what "contradictory" means. And because that is what it means, there cannot be contradictory facts. Take any supposedly contradictory facts, such as A and B, to keep matters simple and to dodge the impossibility of producing a real example. If A is a fact then the statement of this fact, "A," is true. And the same goes for B and "B." But then the statements "A" and "B" are both true, and so not contradictory after all. The very existence of contradictory facts would mean they weren't really contradictory.

The idea that there are contradictions in reality, and not merely in our beliefs about it, is possible only when "contradiction" is used to mean something other than contradiction. Misusing words may normally be dismissed as mere illiteracy. When it is done systematically by people regarded as great thinkers, however, the misuse is liable to gain some currency. And that is what happened with "contradiction" at the hands of the nineteenth-century philosopher Hegel and his Communist successors, Marx, Lenin, and Mao Tse Tung.

According to these Dialectical Materialists, real contradictions aren't just common, they are essential to everything. Here is Mao Tse Tung on the matter:

> Engels said, "Motion itself is a contradiction." Lenin defined the law of the unity of opposites as "the recognition (discovery) of the contradictory, *mutually exclusive*, opposite tendencies in *all* phenomena and processes of nature (*including* mind and society)." Are these ideas correct? Yes, they are. The interdependence of the contradictory aspects present in all things and the struggle between these aspects determine the life of all things and push their development forward. There is nothing that does not contain contradiction; without contradiction nothing would exist.[5]

---

5. Mao Tse Tung, "On Contradiction," *Selected Works* (Peking, Foreign Language Press, 1967), p. 313.

This passage helps us to understand not only what Dialectical Materialists mean by "contradiction," but also what it really means, by providing an example of the latter. Mao begins by agreeing with Lenin that what is contradictory is mutually exclusive. Then he claims that what is contradictory is interdependent. I suppose having a contradictory definition of "contradiction" is no less than you would expect of someone who thinks everything is essentially contradictory.

Insofar as such muddles do not render their interpretation incomprehensible, Dialectical Materialists seem to use "contradictory" simply to mean opposed or conflicting. Only this interpretation makes sense of Lenin and Mao's opinion that the bourgeoisie and the proletariat are contradictory, along with the city and the country, + and −, and everything else besides.

But then Dialectical Materialism provides no excuse for holding contradictory beliefs. Even if Dialectical Materialism were right, it would show only that oppositions or conflicts are rife, not that real contradictions are. You can't show that reality is full of contradiction by calling conflicts "contradictions." No more than you could show that goblins exist by calling birds "goblins."

The only sense in which the world is full of contradiction is that it is full of contradictory opinions and statements. And so it is also full of error. If opinions are contradictory then one of them is false. Contradict yourself and you are sure to be wrong. Not caring about contradiction is the same as not caring about the truth.

# Equivocation

$A$re Christians good?

You might think the tricky word in this question is "good." After all, "good" is a topic of philosophical debate, and there can be no better indication that a word is tricky. But, in fact, the problem lies not with "good" but with "Christian." However challenging it is to define "good," most of us share a sufficiently common understanding of the word to agree in most of its applications. And, more importantly for the purpose of this chapter, it is not ambiguous. There are not two or more clear and distinctly different meanings of the word.

"Christian," however, is ambiguous. It can be used to refer to a person who holds certain beliefs, such as that God created the universe and that Jesus is His son (and also God Himself, if our Christian is a Trinitarian). If this is how "Christian" is understood, then the question "Are Christians good?" is an interesting one. They might be or they might not. To find the answer we will have to look for evidence, such as a lower than average proportion of Christians in prison or higher than average donations to charity or some other such fact.

There is another common use of "Christian," however, on which our question is not in the least interesting, namely, the sense in which "Christian" just means "good." This is employed when people describe immoral acts as "un-Christian," or when Father Ted's congregation responds to the revelation of his pederasty by declaring that he is not a "real Christian" after all. If someone qualifies as a Christian only if he is good, then of course Christians are good—it is true by definition. On this interpretation, it is an open question whether those who believe in the divinity of Jesus tend to be Christians.

This ambiguity is harmless, provided we keep clear about which meaning we are using. Trouble comes when we slip between the two meanings, despite the validity of our argument, requiring us to keep to just one meaning, that is, when we *equivocate*.

Suppose, for example, that Jack recommends Christianity to Jill on the ground that it is the path to virtue. Jill expresses some doubt about this, pointing out that most mafia assassins are Christians. Jack responds that Guido cannot be counted a Christian; no Christian would have whacked the Don's nephew.

Jack has equivocated. He uses "Christian" in its first, belief-based sense when he recommends Christianity as a path to virtue. Then he employs its other sense, on which it is definitionally true that Christians are good, to eliminate an irritating counter-example. Properly, to eliminate Guido as a counter-example, Jack would have to show that he did not believe in the divinity of Jesus—which is not entailed by the fact that he whacked the Don's nephew.

If Jill points this out, Jack is likely to protest that Christianity is more than mere belief in the divinity of Jesus. It also involves a moral code, the ten commandments and all that. Guido clearly broke the code, so it is no cheat to deny him the status of a Christian. Alas, Jack has again changed his definition of "Christian." Now it requires believing in the divinity of Jesus *and* being virtuous (assuming the Christian moral code is correct). And this makes Jack's advice worthless. On this interpretation of "Christian," telling someone who seeks a path to virtue that she should be a Christian is no better than telling someone who seeks a third leg that he should be a tripod.

Jack cannot have it both ways. Either he is making an interesting claim about a means to an end or he is simply defining that end. If the former, then he will have to deliver evidence for his claim. If the latter, then, though he may have eliminated the possibility of wicked Christians, he will have rendered Christianity a badge of honor for those who attain virtue, not a path to it. Either way, Jack must pick one interpretation of "Christian" and stick to it.

## Poverty and Poverty

Jack's equivocation on "Christian" is a trick for evading refutation. But ambiguity can be used for other purposes as well. We all know that boys will be boys. This saying takes advantage of two senses of "boy" to make what superficially looks like a tautology into an informative statement. Boys (human males under a certain age) will be boys (unruly brats who cause endless trouble for everyone around them).

More often, however, ambiguity is used to move, without the help of any supporting argument, from a plain factual claim to one laden with moral evaluation. Two examples—one from the British Labour government and another from their former guiding light Karl Marx—will illustrate the trick.

Shortly after coming to power in 1997, the New Labour government announced the shocking fact that 35 percent of British children live in poverty. In a country as wealthy as Britain, how could this be? Something must be done! Economic policy must change!

Some will believe anything bad, but this one really stretched credulity. The poorest in Britain are the unemployed. They receive free housing, free medical care, free education for their children, and small cash sums to pay for food, clothing, and transportation.[1] Most of the poorest in Britain own or rent telephones, television sets, refrigerators, ovens, stereos, and even cars. The idea that 35 percent of British children live in poverty was literally incredible.

Then it emerged that when the government said "poverty" they meant something quite particular. When *you* hear of poverty, you may have images of families struggling to feed them-

---

1. Housing, health care, and education are never really free. Builders, nurses, and teachers must all be paid, equipment purchased, and someone has to come up with the money. This is often forgotten by those who insist that, for example, education "should be free to everybody." This is simply impossible. Someone must pay, the question is only who. "Not me" is a popular answer, but not everyone can be indulged.

selves, wearing worn-out old clothes, and living in ramshackle and dangerous buildings, unable to send their children to school or have their medical problems treated by a doctor. But this is a rather old-fashioned view and not at all the way a modern political movement like New Labour would approach the issue. No, when New Labour says that someone is a pauper, they mean that his household income is less than 60 percent of the national median household income. Thirty-five percent of children live in poverty because they live in households whose income is less than 60 percent of the national median.

The Labour government does not define "poverty" the way the rest of us do. If that is not already clear, consider the fact that a policy that reduced your income could lift you from poverty, provided it reduced the median income more. Equally, no increase in the incomes of the poor could lift them from poverty unless accompanied by a lesser percentage increase in the median income. Since 1979, the real incomes of all income deciles in Britain, poorest 10 percent to richest 10 percent, have increased. But the proportion living in poverty (Labour style) has increased, because the difference between low and middle incomes has increased.

Labour has introduced a new classification into policy discussion, namely, "household income less than 60 percent of the national median." This classification may be useful for understanding certain social phenomena; perhaps this group suffers peculiarly high rates of criminality or of domestic violence, regardless of absolute income levels. But it is simply misleading to use the term "poverty" for this classification. "Poverty" already has a perfectly well-understood and quite different meaning.

The advantage to Labour in using the term is that poverty, as commonly understood, has powerful evaluative associations that lead us all to the conclusion that "something must be done!" Tell me plainly that 35 percent of British children live in households with incomes less than 60 percent of the national median and I am not at all sure that something must be done, especially if the absolute wealth of these families delivers adequate welfare and opportunity. To get from this statistical observation to the conclusion that policy must be changed requires considerable argument. But argument is irksome and made unnecessary by simply labeling our statistical classification "poverty."[2]

## Exploitation, Mexploitation

New Labour has rejected Old Labour's Marxist ambitions. Private ownership of the means of production is now fine by the Labour party. Indeed, many of its members openly practice private ownership. But those earnest undergraduate days, those hours reading *The Communist Manifesto* and getting halfway down page one of *Capital*, must have left their mark. For New Labour's trick with "poverty" displays a certain Marxist tendency.

Marx's work in economic and political theory broke new ground—a grand theory of human history as the consequence of economic and technological forces. Groundbreaking theories usually introduce new concepts. Genes, schizophrenia, and price equilibria, though now familiar, were once the novel concepts of

---

2. The government's measure of relative poverty is discussed in more detail on pp. 134–137.

biological, psychiatric, and economic theories. So it was with Marx, who posited unfamiliar economic forces and social phenomena for which he needed new terms.

In physics it is common to introduce wholly new words for new theoretical entities—positrons, quarks, and so on. But in the social sciences this is not at all the norm. Usually, familiar words are combined in a new way to label the new concept: marginal utility, Gross Domestic Product (GDP), and so forth. This is what Marx did. Except, instead of employing the normally tedious language of the social sciences, he went for a little excitement. His theoretical concepts get names like "exploitation" and "alienation." For example, in a capitalist system, business owners typically pay their workers less than they receive from selling what the workers produce, since otherwise they won't make a profit. Marx calls this phenomenon the "exploitation" of the workers.

Words like *exploitation* and *alienation*, in ordinary use, have strong negative connotations, but please, set all that aside. These words are used in Marx's theory purely as technical terms to describe certain economic or social phenomena. This is science.

Alas, new stipulative definitions of familiar old words are hard to bear in mind. We can't help but slip back to their ordinary meanings. Though generally unfortunate, this slippage does have one desirable consequence—for Marxists, at least. It delivers, as if by magic, the sought after condemnation of capitalism. For behold, Marx has shown that capitalism inherently involves exploitation and alienation. Who could be in favor of such a dreadful system? Hey Presto!

Marx has, of course, shown nothing of the sort, at least if "exploitation" and "alienation" have their ordinary meanings.

At best, he has shown that capitalism inherently involves Mexploitation and Malientation (I place an "M" at the front of "exploitation" and "alienation" to distinguish Marx's technical terms from their ordinary homonyms). And that, on its own, provides no ground for condemning capitalism. I, for one, am entirely in favor of economic systems where business owners can make a profit. I agree that this involves Mexploitation—how could I not, given the definition of Mexploitation? But I deny it is exploitation.

## Verbal Solutions

The equivocator tries to replace hard intellectual graft with semantic sleight of hand. Capitalism may indeed exploit the workers, but you can't show this simply by spelling profit "exploitation." Redefinition, or slipping between different meanings of the same word, cannot deliver an intellectual free lunch where you arrive at informative conclusions without paying a price in evidence and argument. And, if it won't solve your intellectual problems, it certainly won't solve any practical problems. But playing with words is much easier than tackling reality, and often overwhelmingly seductive for tired policy makers.

The last Conservative government of Britain believed the country would benefit from having a better-educated population. In particular, they wanted more university-educated citizens. Never mind why they wanted this; let us simply accept it as their goal. The problem lay in achieving it. Significantly increasing the number of university graduates would require a great increase in the capacity of Britain's university system—more universities or at least more places at existing universities. This must have been a daunting prospect, both expensive and slow, since it takes many years to train new university lecturers.

Then the Eureka! moment. I don't know if some erstwhile Marxist had somehow slipped into John Major's administration or whether it was just blind brilliance, but someone hit upon the idea: let's call technical colleges "universities." Amazing! Dozens of new universities at a stroke and hardly a penny spent. And so it was that during the 1990s the number of universities in Britain almost doubled.

The problem with this policy is that it has no effect whatsoever on the amount or quality of education going on in the country. So it cannot deliver what was wanted, namely, a better educated population. It delivers only more people with the badge of a university education. It is the same mistake as thinking you can make everyone rich by devaluing your currency. Look how many millionaires there are in Turkey.

You can't change the world just by describing it differently, or replacing nasty old words with nice new ones. If your shithouse stinks, you won't make it smell any better by calling it a public convenience. You need to clean it. And a cripple won't stand and walk because you call him alternatively-abled.

Of course, words can acquire unpleasant connotations over time and only those who wish to endorse those connotations will continue to use the word. "Nigger" is a perfect example. But a prohibition on calling anyone a nigger, or even mentioning the "N" word, will not in itself eliminate racism or improve the lot of African Americans. On the contrary, if racism persists, other names for blacks will pick up the unwanted connotations.

Euphemisms tend to undermine themselves. The fact that you feel in need of a euphemism shows that you have a dim view of what it refers to. And soon enough, if your distaste is shared by others, the euphemism will have lost its power and all the old associations will have reattached themselves. In a country as blighted by racism as America there is always the need for yet another name for blacks. And when it comes to the place we do our number twos, everybody is always looking for the next nice way of referring to it. Many Londoners have recently adopted the American expression "bathroom," the old French euphemisms "toilet" and "lavatory" having lost their cleansing magic.

Euphemisms usually create no serious problems. They display a certain squeamishness but we still know what is meant. Equivocation, however, as illustrated in the "Christian" and other examples, is a serious intellectual aberration. Whether committed knowingly or by accident, it is a kind of cheat. Beware of it when a line of argument seems too good to be true—when, from the mere definitions of words, informative conclusions about reality are drawn or when value-laden conclusions are derived from purely factual premises. For neither kind of argument can possibly be valid. Any appearance of success is sure to be explained by the presence, somewhere in the argument, of an equivocation.

# Begging the Question

In the 1980s, the Auckland University student newspaper ran a weekly column by the university chaplain. "The Chaplain's Chat" was its comforting title, with content to match. In 1984, a new column appeared alongside "The Chaplain's Chat"—"The Egyptian Chaplain's Chat"—in which the author would give a brief account of the characteristics of one of the ancient Egyptian deities, the sun god Amon Ra and his colleagues.

This new column was not well-received by the campus Evangelical Union, a society for the most enthusiastic of Christian students. They wrote a letter to the newspaper demanding the withdrawal of what they considered to be a weekly dose of blasphemy.

The editor did not oblige them. Instead, he wrote an editorial lamenting the Evangelical Union's lack of tolerance. It ended with the slogan: "Believe what you will but tolerate the beliefs of others."

He must have been pleased with this *bon mot*: tolerant, yes, but firm in the face of dangerous fanaticism. Unfortunately, it is impossible to obey this edict when you believe that someone else's beliefs are intolerable. Which is precisely what the outraged evangelical Christians believed.

The editor thought he could settle the dispute between the Christians and his newspaper without engaging in theological debate. A general appeal to tolerance would do the trick. But he did not really avoid taking a theological position. You would tolerate "The Egyptian Chaplain's Chat" only if you disagreed with the evangelical Christians about this kind of Egyptian chatting leading to the everlasting flames of hell. Far from answering the evangelical Christians' objection, the editor simply assumed they were wrong about this; his plea for tolerance *begged the question*.

The fallacy of begging the question consists in taking for granted precisely what is in dispute, in passing off as an argument what is really no more than an assertion of your position.

Another explanatory example will be useful, since this fallacy is not as immediately obvious as those discussed in earlier chapters.

Suppose Jack is a Libertarian of the variety who thinks that government policy must respect the absolute property rights of individuals over their income and other possessions, no matter what the consequences for aggregate welfare in society. He thinks there should be no tax or redistribution of wealth and that minimal state functions—justice and defense—should be funded by a state lottery, fines on criminals, and booty from military adventures.

If Jill is like most opponents of this view, she will protest that Jack's "no tax" policy would lead to mass poverty. By so arguing, she assumes precisely what Jack denies, namely, that such consequences warrant the violation of property rights. In other words, Jill's objection begs the question. She takes for granted what is in dispute. To refute Jack's "no tax" policy, Jill will have to show that property rights are not really absolute, that certain consequences, such as mass poverty, can warrant overriding them.

Begging the question occurs when people fail to get to the root of their disagreement. This may explain its popularity. Getting to the root of a disagreement will often force you to scrutinize your fundamental assumptions, because that is often where the disagreement lies. But this can be an unpleasant business. These fundamental assumptions will normally have been acquired without even a moment's thought and seriously thinking about them might well bring on a nasty fit of the ideological wobbles. Why do you believe that people have absolute property rights? Or that they don't? Most people would rather have teeth pulled than seriously confront such questions. So they don't. They simply take their favored answer for granted and talk straight past those who disagree.

Whatever its cause, the ubiquity of question-begging means that it usually goes unnoticed. But look hard and you will see it going on in almost all debates. The examples to be discussed here are but the tip of a massive iceberg.

## Tolerance

Begging the question is especially common in disputes about prohibition, on both sides. As in the example of the Egyptian chaplain, those on the liberal side often appeal to some general principle of tolerance, despite the position of their prohibitionist adversaries being precisely that the behavior at issue is intolerable. The most egregious example occurs in the abortion debate, where a common riposte to those who would criminalize it runs, "If you believe abortion is wrong, that's fine, don't abort your pregnancies. But show tolerance toward others who don't share your beliefs." Surely everyone, whatever their personal views, ought to be pro-choice.

Anyone who has ever actually listened to an anti-abortionist ought to be incapable of this response. Anti-abortionists don't think abortion is some kind of lifestyle *faux pas*, like serving white wine with beef or driving a 4x4 despite never having seen the countryside. They think it's murder. They think killing a fetus is no different morally from killing an adult. If their prohibitionism is wrong, it is not because they are insufficiently tolerant of murder. It is because killing a fetus is not really murder.

Tolerance is irrelevant in the abortion debate. If abortion isn't murder, toleration isn't required; if it is murder, tolerating it would be a vice.

Everyone favors tolerance—but only, of course, of what should be tolerated. This qualification is the tricky bit; it is where disagreements tend to arise. And when they do, extolling the virtue of tolerance is of no help, because it can't tell us what should be tolerated and what not. Like most preaching, it's as empty as it is pompous.

## Intolerance

It is not only the liberal side that tends to beg the question in prohibition debates. Consider the drugs debate. Why, for example, should snorting cocaine be illegal? A common answer is that if it were legal there would be more of it.

Well, quite! That is precisely why those who would legalize it would legalize it. They think that people who want to snort cocaine ought to jolly well get on and snort cocaine without being stopped by the state. To protest that legalization would lead to more cocaine-snorting assumes what is in dispute: namely, that there is some reason why people who want to snort shouldn't.

"There'd be more of it" is a popular prohibitionist argument, no matter what the activity in question. I have heard it employed against almost every victimless crime, from drug-taking to consensual adult incest. But, it is a perverse argument, because it draws attention to what you'd think counted on the liberal side, namely, that people would like to indulge more. It is incumbent on the prohibitionist to show why they shouldn't. Otherwise the objection simply begs the question.

The most common argument of prohibitionists, who feel the need to show why people should not indulge, is to claim that the behavior is bad for you; the crime is not really victimless. This argument is usually question-begging, too, since liberalizers tend to agree with John Stuart Mill that nothing should be prohibited on the ground that it harms the voluntary participant: harming others should always be required for prohibition.[1] But, even if we accept the paternalism of our prohibitionists, their argument still ends up begging the crucial question.

Consider again cocaine. Let us grant all the harm done by snorting cocaine. It can damage the inside of your nose; after a

---

1. John Stuart Mill, *On Liberty*, reprinted in *Utilitarianism* (Glasgow, Fount, 1978), pp. 126–250.

hard night on the stuff you are apt to feel somewhat tired and grumpy the next morning; and if you become a heavy, habitual user your productivity at work and intimate relationships may suffer. These are certainly costs of using cocaine, not to mention the $100 per gram. But if you seek to ban something on paternalistic grounds, you must consider not only its costs but also its benefits.

Eating, for example, involves effort and risk. But no paternalist would consider banning it because its benefits outweigh these costs. Proper paternalistic prohibition requires not merely a list of the costs but a cost-benefit analysis. The costs must be set against and exceed the benefits. The banned activity must be a net cost.

The main benefit of snorting cocaine, perhaps the only benefit, is the pleasure it gives the snorter. Prohibitionists never consider this benefit. They will acknowledge some benefits of legalization, such as removing the criminal element from drug-trading, but they never consider the increase in pleasure caused by increased cocaine-snorting. Yet that is the whole point of legalizing it. If you don't count what makes people want to do something, the cons will always outweigh the pros. Consider kissing. If you set aside the fact that people enjoy it, then there isn't much to be said in its favor—it is just a good way of spreading germs. Ban it!

The paternalist who thinks cocaine use is a net cost tacitly assumes that the pleasure is insufficient to compensate for the costs. Since what he assumes here is precisely what is at issue in the debate, he begs the question. Worse. His assumption is sure to be wrong. Those who prefer to snort cocaine after considering all the costs must value the pleasure more. Otherwise they wouldn't prefer to snort. According to their values, snorting cocaine is a net benefit.

## Begging Political Questions

Many politicians like to present themselves as pragmatists. Theory and ideology are not for them. They are in the business of making practical changes that improve people's lives. When it comes to making a policy decision, they are guided by common sense and a thorough command of the nuts and bolts.

Alas, it is harder to avoid ideology than these pragmatists would have us believe. What, for example, does it mean to improve someone's life? That they are richer? That they have more free time? That they are more likely to go to heaven when they die? If you can't answer this question, how can you know whether any policy represents a practical measure to improve people's lives?

Nor is it just the ends of policy that require ideology or theory. So does knowing the means to those ends. Will free trade enrich the people of America or will protectionism? To answer this question, you need to engage in economic theory. For some, it is just obvious that free trade will make Americans poorer by exporting jobs to low-wage economies. But this isn't obvious. In fact, it is false. (Any introductory economics textbook will explain why.)

Whether you like it or not, politics is thoroughly ideological. At the heart of most debates about a specific policy lies a more general ideological disagreement. Yet our pragmatic politicians will not engage with the general question. When arguing for or against a policy, they simply take their ideology for granted, perhaps unaware that they are doing so. And so they beg the question. They take for granted precisely what is at issue in the debate.

Take a topical example. Was the United Kingdom government right to increase income tax to spend more on the National Health Service (NHS), as they recently did? Their justification is that it will lead to more patients being treated in NHS hospitals. This outcome justifies the policy, however, only if the NHS is the most efficient way of delivering health care. Yet that is precisely what opponents of the policy dispute. The debate is really about the relative efficiency of state and private-sector health-care provision. The government's defense of their policy begs the question because it simply assumes that the NHS is more efficient.

The desire to avoid ideological debate can lead to even worse crimes than begging the question. Often the result is total incoherence. Consider, for example, the current debate with Islamic fundamentalism. Most Western politicians reject Islamic theocracy and sharia law. They have their various grounds for disapproval: it is insufficiently democratic, sharia law fails to protect women's rights, and so on. These may be good objections, but they suffice for the rejection of Islamic theocracy only if the basic religious beliefs of Islamic fundamentalists are wrong. If Allah really does demand sharia law, as the fundamentalists claim, then we should all adopt it as soon as possible. Despotism and sexism are well worth it if they will save us from the everlasting flames of hell.

Of course, most Western politicians do reject this Islamic idea about Allah's will. But they won't say so in public. When arguing against sharia law they never point out that Allah does not really exist and so can't really have insisted on any form of law, sharia included. On the contrary, most who engage in the debate will spend quite some time professing to have enormous respect for Islam. But if they do respect Islam why will they not adhere to its political prescriptions?

Perhaps I misunderstand what politicians mean by "respect." Perhaps you can respect a religion whose basic tenets you think false and whose political ideology you believe to be despotic and sexist. Even so, you must admit that it is a little confusing, when rejecting Islamic theocracy, to focus on your respect for Islam rather than your belief that its basic ideas are false.

A politician cannot tackle directly every theoretical challenge to his position. He hasn't the time, and many of them are too silly to warrant the effort. He must pick his fights, normally, against his main political rivals. But, once he picks his fight he must fight properly, trying to land blows above his opponent's belt rather than just dancing evasively around the ring waving triumphantly to his fans in the crowd.

## Disguised Assumptions

Apparently plain statements can embody contentious assumptions. For example, describing an income tax cut as a "giveaway" assumes that a citizen's gross income is not her own but is, rather, the property of the government. Describing the government's spending plans as generous embodies the same assumption. The virtue of generosity does not consist in giving away others' money: it requires you to give away your own.

It might be true that everyone's gross income is the government's property, but if you simply assume it in a debate with someone who disagrees, you beg the question. Which means that you also beg the question by using language that implicitly assumes it, like describing government spending as generous.

It isn't only statements that can in this disguised way beg the question. Questions can too. "Have you stopped beating your wife?" is the famous example. Answer yes or answer no and you seem to confirm your status as a wife-beater. But if your wife-beating status is in dispute, the question begs the question.

Even a name can beg the question. The Peace Movement of Cold War days provides a nice example. The Peace Movement consisted of those in favor of nuclear disarmament, even unilateral disarmament on the part of the West. They thought this would promote peace and save the world from nuclear Armageddon. Their opponents believed in nuclear deterrence. They thought this would promote peace and save the world from nuclear Armageddon. It is presumptuous, you will admit, for one side in this dispute to label itself the Peace Movement.

It takes a terrible pedant to worry about such contentious built-in assumptions, and pedantry has got itself a bad name. But don't let that put you off. As Bertrand Russell said, a pedant is just someone who prefers his opinions to be true.

# Coincidence

My day job brings me into frequent contact with the financial controllers of banks. Quite dull, you probably think.
Usually. But occasionally, they say remarkable things. Only last
month I made the mistake of suggesting to one that perhaps the
events we were discussing had been merely coincidental. He
regarded me with an expression between pity and contempt and
declared, "I don't believe in coincidences."

Which is interesting. Here is a controller, his job to manage
the financial security of a bank, and he declares that he doesn't
believe in coincidences. It's like the chief of police saying he
doesn't believe in burglary, or a firefighter expressing her doubts
about the reality of smoke.

Any bank faces risks to its financial security. Those to whom
it has lent money might fail to repay it, the value of bonds and
shares it owns might collapse, or an employee might lose it a fortune through "rogue trading," as Nick Leeson did Barings Bank.

And any combination of such events might, purely by coincidence, come at the same time, compounding the loss. Since a financial controller's job includes calculating how much capital his bank must hold as insurance against such coincidences, it is best if he believes in them.

Not believing in coincidences is part of a manly pose adopted by many who fancy themselves savvy. They don't trust strangers, they don't count their money while sitting at the table, and they don't believe in coincidences. It's a shame, because coincidences happen all the time; they are statistically guaranteed to. And failing to believe in them makes you believe in things that really don't exist, those forces imagined to explain away the appearance of coincidence.

This chapter concerns the mistakes people make when they fail to recognize simple coincidence at work, from paying their staff more than they should to believing in God.

## What You'd Expect

In cricket, batsmen most frequently get out on the score zero, though it is quite possible for them to score more than a hundred runs in an inning. I once heard a panel of professional cricketers attempting to explain this fact. Two explanations were popular. One was that when a batsman has just begun his innings, and so has a score of zero, he hasn't had time to "get his eye in." The other was that batsmen are nervous until they have scored some runs. Both explanations attempt to show why a delivery (the equivalent of a pitch in baseball) is more likely to get a batsman out when he is on zero than when he is on a greater score.

But this is not required to explain why zero is the score on which batsmen most often get out. The way cricket is scored means that, even if each delivery a batsman faces is equally likely to get him out, regardless of his score, zero would still be the score on which he would most often get out. A batsman always begins on zero and each time he faces a delivery he can either get out or score anything between zero and six runs. This means that zero is the score on which batsmen face the most deliveries. So you should *expect* zero also to be the score on which they will most often get out. No need to postulate any further explanation in terms of acclimatization, nervousness, or anything else.

Something beyond the way the game is scored would be required to explain this statistic only if the frequency of batsmen getting out on zero were greater than would be expected from the game's scoring system alone.

Lessons from cricket are generally applicable to life. What, if anything, is required to explain some statistical fact depends on what you'd expect in the first place—what you'd expect if everything were playing out according to the odds. This is so not only when we consider a typical or average outcome, such as the score on which batsmen are most often out, but also when we consider extraordinary events. Extraordinary events are just what you should expect to happen, at least occasionally.

Consider the bond traders who work for investment banks. Some of them simply execute orders to buy and sell bonds on behalf of their clients, but others trade on behalf of the bank's own account, proprietary traders, as they are known. Some of these proprietary traders make vast sums of money for the banks that employ them. These are the "star traders" whom banks compete fiercely to employ, tempting them with massive salaries and bonuses. Yet there is no reason to believe that these people possess any skill worth paying for. Their extraordinarily profitable trading, on which they are judged stars, may be nothing but luck.

Before going into this matter further, it will be useful to consider the card game Chancy. The rules of Chancy are simple. Spread before you is a well-shuffled deck of cards, face down. You play by drawing a card from anywhere in the deck. If it is a seven you earn no points. If it is above seven, then you earn the value of the card less seven (e.g., a Queen earns you five). If the card is below seven then you lose seven less the value of the drawn card (e.g., draw five and you lose two). After each turn, the card drawn is returned to the deck, which is then reshuffled. The winner is the player with the highest score after an agreed number of draws by each player.

Over the long run, you should expect your average score to tend toward zero, since the likelihood and size of earnings and losses on each draw is equal. But there is no impossibility in having a run of luck and accumulating a large positive score. On the contrary, the probability of any such run of luck can be calculated.

For example, the probability of five consecutive draws being earners (i.e., above seven) is 2 percent, and the chance of five consecutive draws being *big* earners (Jack or better) is 0.065 percent. This means that if a large number of people sat down to play a five-draw game of Chancy, you would expect 2 percent of them to draw only earners and 0.065 percent of them (1 in 1,500) to draw only big earners. These winners possess no skill lacked by the other players; the game provides no way for skill to affect the outcome. The winners have just been lucky, as we knew in advance some would be. Just as the rules of Chancy guarantee that, in a large enough sample, the average score will be zero, so they also guarantee that some players will be big winners.

Now, let's return to bond trading. How much it resembles Chancy is hotly debated. Those traders with "high scores" will assure you it is a game of skill, not luck. But that is irrelevant for the point I wish to make. Even if bond trading is a game of skill, you can still win by luck. The prices of bonds go up and down, by lesser and greater amounts. Even if a trader made her buy and sell decisions by tossing a coin, given enough luck she could still be a big winner.

Suppose, then, that you own or manage an investment bank. One of your proprietary traders has five consecutive big wins and earns your bank a fortune. Deutsche Bank hears of this incredible feat and attempts to poach him from you, offering a salary of $750,000 and guaranteed minimum bonus of $2,000,000. Should you top this offer to keep your star trader or cheerfully say goodbye to someone who has just had a run of luck that there is therefore no reason to believe will be repeated?

There are two ways to answer this *skill or luck* question. One was suggested already. If it is skill rather than luck, then this successful trading should persist. Observe your trader at work. If his success continues over very many trades and through different market conditions (trending, volatile, and so forth), then the probability that it is just luck will approach zero: he is clearly a wizard of finance.

The second is to think about what information is available to your trader and how he makes his decisions. Could he possibly predict bond price changes or follow a trading strategy that beats the odds?

Answering the *skill or luck* question is not easy. The expected random distribution of performance in bond trading is harder to calculate than the distribution of Chancy scores; the bond markets are more complex than this simple card game. So it will be correspondingly harder to tell whether or not a trader's performance can reasonably be attributed to luck. And this complexity also makes it difficult to know whether or not a trader's decision-making processes could possibly give him any advantage. Yet, if you manage an investment bank, you should try to answer these difficult questions. Otherwise, for all you know, you are paying people enormous salaries merely for having been lucky.

Remarkably, investment banks make no serious attempt to answer the *skill or luck* question. They simply assume that a run of success is evidence of financial wizardry, even when the success would require only moderate luck. When the sums involved are large enough, even one or two wins will do. "Look, she has just made us $500 million. That can't be luck!" This is precisely what was said about several traders of emerging markets government bonds in the late 1990s—only days before they lost their banks millions of dollars when the Russian government defaulted.

The strongest evidence that investment banks just assume success is due to skill is their habit of offering successful traders from competitor banks vast sums of money to switch employer. The tactics of these traders are a mystery to the banks that court them. Perhaps they bought when the coin landed heads and sold when tails. Their new employer neither knows nor, apparently, cares.

Investment bankers seem to agree with those primitive tribespeople who assume that any good or ill fortune that befalls someone must be due to some quality of the person—that there is really no such thing as luck at all. In this spirit, I suggest a new line of business for them. Investment banks should employ former winners of Lotto jackpots to buy Lotto tickets for them. No salary could be too high for these wizards of ticket selection.

## Coincidental Healing

I have a friend who believes in homeopathy. He once tried to convince me that its effects are not due to the placebo effect by telling me about a sick dog that responded positively to homeopathic potions. It's not a bad argument regarding the placebo effects of homeopathy. Since dogs don't know they've been given medicine, improvements in their health can't be due to the psychosomatic effects of such knowledge. But my friend had mistaken my concern. I don't think homeopathy works via the placebo effect; I don't think it works at all.

How can I maintain this skepticism in the face of the cured dog? And in the face of my friend, who has had the same ailment cured several times by homeopathy? Quite easily. This dog and my friend would have got better anyway. Consumption of homeopathic medicine preceded their recoveries but that doesn't mean it caused them. It's just a coincidence.

To think that what follows something must be caused by it is to commit a fallacy so famous that it has a Latin name: the *post hoc ergo propter hoc* fallacy (the *after this therefore because of this* fallacy). Events may happen in sequence, and even in close proximity to each other, without being cause and effect. Just as I sit down, the lightbulb blows. It doesn't mean that sitting down caused the lightbulb to blow.

Event A caused event B only if B's occurrence depended on A's: only, that is, if B wouldn't have occurred if A hadn't. This condition isn't met in the case of the lightbulb. The lightbulb's blowing didn't depend on my sitting; it would have blown even if I hadn't sat down. B can follow A without depending on A.

No one would think that sitting caused the bulb to blow because everyone knows that isn't how the world works; sitting isn't the kind of thing that causes lightbulbs to blow. Knowing how the world works—knowing what sorts of things cause what—depends on what we observe following what, but not on a mere handful of observations. One dog getting better after taking a homeopathic medicine cannot alone show that homeopathy has its alleged curative effects.

How then do we confirm such causal claims?

First, let's take a probabilistic view of causation. That is, let's say not that if A hadn't happened B wouldn't have, but only that if A hadn't happened the chance of B happening would have been lower. In other words, causes make their effects more likely. This is the sense in which smoking causes cancer and regularly playing Lotto causes penury.[1]

Now, suppose our causal hypothesis is that taking homeopathic medicine cures influenza. The chance of recovery in any period of time is higher if you take a homeopathic medicine than

---

1. The idea that causation is probabilistic is now widely accepted by scientists and philosophers. Though not essential to the case for probabilistic causation, the probabilistic nature of quantum physics has convinced many.

if you do not. We test this hypothesis by taking two sufficiently large groups of influenza sufferers who are in all identifiable respects the same: each group has the same distribution of ages, sexes, races, general health, and so forth. To one group, we give the homeopathic medicine (the test group), to the other, we do not (the control group). We then check the recovery times of the people in each group.

In the group who have taken no medicine, recovery times will vary. Some will get better remarkably quickly, most in about a week (let's say) and some will take several weeks to recover.

**Recovery Times in the Control Group**

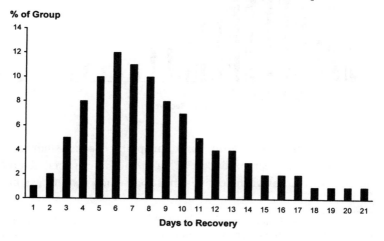

This immediately shows why individual cases of fast recovery prove nothing about the efficacy of homeopathic medicines: some people will recover quickly with no medicine at all. To establish homeopathy's efficacy the distribution of recovery times in the group given the medicine must be significantly better than that in the control group, as illustrated in the following chart.

**Recovery Times: Test Group vs. Control Group**

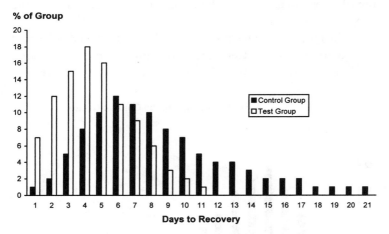

I do not believe the claims for homeopathy because such test results have not been produced.[2] There is nothing special about homeopathy here. All medicines must pass such tests to estab-

---

2. Jacques Benveniste and colleagues published a paper in *Nature*, 1988 reporting experimental results supporting homeopathy. However, these results have not been replicable in subsequent experiments (see Forbes et al. in *Nature*, 1993). The standard of proof for homeopathy should be especially high, since it asks us to believe something incredible: that the effects of water depend not on what is in it, but on what used to be. (See pp. 93–94.)

lish their efficacy. Indeed, there is nothing special about medicine. Any true claim about what causes what should be able to pass such tests.

Believers in dubious medical treatments have a ready answer for negative test results. "Of course," they will tell you, "it doesn't work for everyone." This response can be legitimate, provided the type of person for whom it supposedly works is specified. We can then test this hypothesis by repeating our experiment, except this time with our two groups populated only by the type for whom the treatment allegedly works. If the hypothesis is true, we should see a better distribution of recovery in the test group than in the control group.

Normally, however, the only criterion suggested for someone being of the right sort is that she does, in fact, get better. But then we are back where we started. People take the medicine, some get better quickly while others do not, and we have no reason to believe that everyone would not have done equally well without the medicine.

This simplified account of how alleged medical effects are tested is not controversial. Yet many refuse to take it seriously where their own health is concerned. Documentaries about alleged medical scandals are never short of people who claim to know that their cancer was caused by the power pylons in the backyard or that their daughter's suicide was caused by the antidepression drug she was taking, even when experiments of the kind I have described show no connection between power pylons and cancer or antidepressants and suicide.

The journalist nods encouragement as the cancer victim assures us that he knows the cause of his disease. But how he acquired this knowledge, contrary to the results of scientific experiments, is never explained. How does having cancer bestow a magical power to discern its cause? All he knows is that there is a power pylon in the backyard and that he has cancer. Which isn't enough to know that the pylon caused the cancer.

## Thank You Lord for Making Me Likely

The existence of each of us is extraordinarily improbable. Had things been even slightly different—had that shrapnel entered your grandfather's chest two inches to the right, had that train not been delayed so that your parents did not meet at the platform, had the drugstore not sold out of condoms that night— then you would not have existed. How easily all this might never have been. Sets you to thinking, doesn't it?

Indeed it does. And what many conclude is that it could not have been a coincidence after all. Surely nothing that important— my existence—can have been a matter of chance. No, it was fated that my parents should meet at the station, that the drugstore should have a run on Trojans, and that sperm number 203,114 out of the millions issued by my father that fateful night should be the one to win the race and fertilize my mother's egg.

The conceit of this thinking is breathtaking. Had things gone differently—had your mother not met your father at the platform, for example—then although *you* would not have existed, some- one else would have. Your mother would have met someone else and reproduced with him instead. And had things gone differently in other ways, then yet other people who do not now exist would have: even a different sperm from your father would have meant a different person. To say that your existence is fated is to say that fate or God or whoever is supposed to arrange these things prefers you to all the many possible people whom your existence has ruled out. You must really be special.

Most theologians are humble to the point of ostentation and so wouldn't dream of adopting such an egocentric view of how they came to be. But many do attempt to use the improbability of human existence, if not their own personal human existence, to show that God exists. The argument follows just the line above. They begin by noting how improbable the existence of human beings would be if God did not exist. And from this fact alone they conclude that God exists.

George Schlesinger makes the argument as follows:

> In the last few decades a tantalizingly great number of exceedingly rare coincidences, vital for the existence of a minimally stable universe and without which no form of life could exist anywhere, have been discovered . . . The hypothesis that [the requirements for life were] produced by a Being interested in sentient organic systems adequately explains this otherwise inexplicably astonishing fact.[3]

The existence of humans is indeed improbable. The laws of nature that govern our universe are but one set out of infinitely many possible sets of laws of nature. And had they differed only slightly the universe would be a mere swirl of subatomic particles, free from medium-sized objects like rocks, trees, and

---

3. George Schlesinger, *New Perspectives On Old Time Religion* (Oxford University Press, 1988), pp. 130, 133.

humans. And even given the actual laws of nature, evolutionary history could have taken different twists and turns and failed to deliver human beings.

So the premise of this theological argument, that human existence is unlikely, is true. But it provides no ground for believing in God. Showing why not is useful because it illuminates a common fallacy in probabilistic reasoning. Before identifying this fallacy, however, it is easy to see that the argument is invalid. If it were valid we could conclude, without the aid of any evidence, that all lotteries are rigged. Which, I hope you will agree, we cannot.

Suppose Jill has won Lotto. This was very improbable, about a one in fifteen million chance. Unless, of course, the lottery was rigged in her favor. Therefore, the lottery was rigged in her favor. Or, as George Schlesinger would put it, the hypothesis that the lottery was rigged in her favor adequately explains the otherwise inexplicably astonishing fact that Jill won.

There is nothing special about Jill. Suppose Jack had won the Lotto instead. This, too, would have been very improbable had the lottery not been rigged in his favor. So, if Jack wins, we may also conclude that the lottery was rigged. Indeed, whoever wins, we may conclude that the lottery was rigged in his favor, because his winning would otherwise be very improbable.

The original theological version of the argument has precisely the same absurd consequence. Suppose the laws of nature had indeed been slightly different so that humans did not exist. Then other things would have existed instead. And their existence would have been no less improbable than ours, since if the laws of nature had been slightly different, these things would not have existed. Unless, of course, God wanted them to. So, no matter what the laws of nature and hence what kinds of things in the universe, their improbability would always lead to the conclusion that God exists.

The basic error in the argument is a confusion about probabilities. "Which hypothesis is more probable?," we are asked, "that humans exist by chance or that God made it so?" As a general principle, we ought to believe the more probable hypothesis. And which hypothesis makes our existence more probable? Obviously, that God brought us about on purpose.

If you haven't already noticed the slip, consider another case. Jack has just won a game of poker. Which hand makes this outcome most probable? A royal flush. If Jack had a royal flush, then he was certain to win. So, should we conclude from the fact that he won that Jack had a royal flush?

Obviously not. The probability of getting a royal flush is very low, and winning with a lesser hand is quite likely. The hand Jack most probably won with is not a royal flush, despite the fact that it is the hand that, if he had it, would give his winning the highest probability.

The fallacy in our theological argument should now be clear. The statement,

1. Given God's existence, the existence of humans is probable,

does not entail,

2. Given the existence of humans, God's existence is probable.

To infer 2. from 1. is to make the same mistake as inferring from the fact that a royal flush makes winning most likely, that Jack's winning makes it most likely that he had a royal flush. So this theological argument is structurally defective.

Nor would it work even if we were ready to commit the required statistical fallacy. What makes theologians think that God's existence would make human existence probable, in the way that a royal flush makes victory in poker probable? There's God, sitting wherever He sits, contemplating all the universes He might create, with their various laws of nature, planets, creatures, and all manner of things that we cannot imagine . . . and He chooses this one! I do not mean to indulge in the kind of self-flagellation that Christians enjoy when not trying on this theological argument, but I mean to say can this really be God's best option?

Even given the existence of God, it is entirely a matter of luck that we should exist, because it is entirely a matter of luck that He should prefer this kind of universe. How do theologians explain this extraordinary coincidence? Perhaps God was created by an *über*-god who prefers gods who prefer this kind of universe. But that would be a matter of luck too. And so we will soon have an infinity of gods, each explaining the next god's extraordinarily unlikely preferences.

I have met several people who, when explaining the extreme youth or old age of their parents, have told me, "Of course, I was an accident." Well, if they can admit it, why can't we all? Our existence is not due to the preferences of some fabulous Being: it's just dumb luck. Why people should feel bothered by this I don't know. They have won the lottery of life!

# Shocking Statistics

Statistics are the chemical weapons of persuasion. All good politicians and businesspeople know this. Release a few statistics into the discussion and the effects will be visible within moments: eyes glaze over, jaws slacken, and soon everyone will be nodding in agreement. You can't argue with the numbers.

Yes you can. Even when the numbers are right, they often don't show what they are alleged to. For example, newspaper editorialists are always leaping from statistics about changing behavior to conclusions about changing, and usually worsening, values. Behavior can change, however, not because values do but because the circumstances do. Teenagers commit more street crime now than in 1980. Is that because teenagers have less respect for private property or because there is now so much more to steal from people on the street—most notably, their mobile phones? People eat more now than they did in 1950. Is that because we have become gluttonous or because food is cheaper?

Drawing the wrong conclusion from statistics is an interesting mistake, however, only if the statistics are right in the first place. And often they are not. Consider the following statistics:

  35 percent of British children live in poverty.
  50 percent of small-business owners would switch banks to receive a discount of 0.25 percent interest on their overdrafts.
  25 percent of young drug users have smoked cannabis with a parent.
  2 percent of young women suffer from anorexia nervosa, and 20 percent of sufferers die from it.

Each is from a reputable source. Each is also the result of a simple error of statistical method. (Sources don't seem to become reputable by being good at statistics.)

Understanding these errors is not difficult, as I hope this chapter will show, but it is important. Widespread statistical naïveté allows nonsense statistics like these to become the "hard facts" that inform decision-making.

## British Poverty

Soon after coming to power in 1997, the New Labour government drew our attention to a shocking fact: 35 percent of children in the United Kingdom live in poverty. Not *absolute* poverty, of course; even the poorest are at no serious risk of going without food, housing, schooling, or medical care. Rather, 35 percent of children live in *relative* poverty: by the standards of modern Britain, they are relatively poor.

On pages 99–102, I complained that the Labour government played fast and loose with this ambiguity in the word *poverty*. "We need to fight poverty," they claimed. Why? Because poverty is dreadful and there is so much of it. But this is merely a play on words. Absolute poverty is dreadful (but rare); relative poverty is common (but not so dreadful).

In this chapter, however, I want to set that issue aside and examine only the claim that 35 percent of British children live in relative poverty. This claim illustrates a common way in which statistics can mislead: by being based on an improper measure of the phenomenon in question.

The government measures the number of people who live in relative poverty as the number living in households with incomes less than 60 percent of the national median income. We must accept that 35 percent of children live in such households. Still, why should we conclude that 35 percent live in relative poverty? Why, in other words, is *household income less than 60 percent of the national median* a good measure of relative poverty?

The short answer is that it isn't. In a country like the United Kingdom, disposable income inequality is a hopeless way of measuring relative poverty.

To see this, consider two twelve-year-old boys who live next door to each other. They live in the same quality of house, attend the same school, go to the same doctor when they are sick, wear the same brand of athletic shoes, and so on. Indeed, their material well-being differs in only one respect:

Jimmy's parents give him £10 a week in pocket money, Timmy gets only £5 pounds from his. Should we conclude that, since his disposable income is only half of Jimmy's, Timmy is a pauper relative to Jimmy?

Obviously not. Jimmy and Timmy's *consumption* is almost identical. Let's suppose that the housing, clothes, schooling, medical care, and so on that they both receive are worth £100 per week, and that both spend all of their pocket money. Then Jimmy consumes £110 per week and Timmy consumes £105 per week. Though Jimmy's disposable income is double Timmy's, he is only 5 percent better off.

When a large percentage of consumption is not paid for out of disposable income, differences in disposable income will always exaggerate differences in the ability to consume. And it is the ability to consume that is important with regard to poverty, including relative poverty.

So the government's measure of relative household poverty is wrong. Like Jimmy and Timmy, British households need not pay for much of what they consume out of their disposable incomes. Most importantly, medical care and education are delivered by the state, funded out of tax revenues. And, so far as the government's measure of poverty is concerned, housing is free too, since it uses disposable income *after housing costs*.

In a paternalistic society like Britain differences in disposable income will overstate differences in consumption capacity and hence in the number suffering relative poverty. This point has nothing to do with redistribution of wealth. If taxes were high but all benefits were paid in cash rather than state services, then disposable income would accurately reflect consumption capacity, and relative income would be a reasonable basis for evaluating relative poverty. The further a society moves from the "all cash" model toward an "all state services" model, the worse is the disposable income measure of poverty. And Britain is very far from the "all cash" model.

When presented with a statistic involving something hard to measure, such as poverty, happiness, or beauty, you should always check the measure used. Often it will be a crude approximation, acceptable for some purposes but not others. Sometimes it will be plain wrong.

Having alerted you to the danger, however, I can offer no general guidance on how to tell good measures from bad. Each measure must be examined as it is encountered. This will often be difficult, since alleged statistical facts are usually served up plain by newspapers, politicians, and businesspeople, with little information about the precise measure used. Then the proper attitude is open-minded skepticism.

## Switching Banks and Other Lies

The higher you price the products you sell, the greater the profit you make on each sale (unit profit). So why not just set outrageously high prices? Because you would have no sales. Unlike unit profit, sales volumes decrease as prices rise. If you want to maximize your aggregate profit, as most business owners do, the best price is the one that finds the right trade-off between unit profits and sales volume.

To work out this best price, you need to know the unit profit at any given price and the volume at that price. The former is simple when you know your costs.[1] But knowing how price affects volume requires you to understand the price sensitivity of customers. And that is more difficult.

Experimenting can be dangerous. You might guess wrong and end up losing all your customers or giving away unit profits without gaining volume, which is why companies often conduct market research before making any price changes. Alas, such research often gives misleading results, for a simple reason: people lie. Specifically, they claim to be more price sensitive than they really are.

I recently commissioned a survey of the managers of small businesses in Holland regarding the size of discount required to make them change banks. "How likely would you be to switch to a bank offering a rate of interest on your overdraft 0.25 percent lower than your current bank? Certain, very likely, maybe, very unlikely, certainly not? What about 0.5 percent . . ."

If you took the results of the survey at face value, even the slightest discount would have most Dutch small business managers switching banks in an instant. But small discounts *are* available from some Dutch banks, who do not in fact experience long lines of small businesspeople wanting to open accounts.

The good reason managers don't switch banks for small discounts is that switching banks costs more in time and bother than the discount is worth. On a $20,000 overdraft, 0.25 percent is only $50 a year, and changing banks is a big hassle.

---

1. This is a simplification. Where a portion of costs is fixed (i.e., does not vary with sales volume, such as television advertizing), unit profit also depends on volume, since average unit costs will vary inversely with volume. Most businesses have some fixed costs, so knowing unit profit at any given price also requires knowing volumes at that price. But the point still holds: the difficult part is knowing how price affects volume.

So, why do they say they would switch? I can't be sure. But my guess is that they like to think of themselves as astute businesspeople who would not pass up opportunities for a better deal. And saying you would switch banks involves none of the time or hassle of actually doing it.

It is generally best to be skeptical about the results of surveys that get their data merely by asking people about their inclinations or habits. People have all sorts of reasons to misrepresent themselves. They usually don't mean to deceive, but even if they are only lying to themselves the results will still be unreliable. If you want to know about the sexual prowess of men, for example, I wouldn't advise gathering your information just by asking them.

It is difficult to know in advance what people will misrepresent. For example, you would think that voting intentions are something on which you could take anyone's word. But they aren't. The U.K. Conservative Party's 1992 general election victory came as a surprise to opinion polling organizations, most of which had forecast a comfortable victory for Labour. Their post-election analysis of how they got it so wrong revealed that many people who vote Conservative are reluctant to admit it, even in an anonymous poll. So, be warned. If even Tories can't be relied upon secretly to admit it, there is little you can take at face value.

## Dope with Dad?

It is always refreshing to discover a good news story in the paper. I thought I had encountered one in the *London Times* (Feb. 24, 2003, p. 2) under the headline "Drug Parents." It announced that "nearly a quarter of young drug users have smoked cannabis with a parent." Family life is not dead in Britain after all.

Alas, I read on and discovered that the statistic couldn't be trusted. It was the outcome of "a survey completed by 493 readers of rave magazine *Mixmag*." You will see the problem. Even if those who complete surveys in *Mixmag* can be relied upon to tell the truth about their drug-taking habits, they are hardly a representative sample of young drug users. They are, for a start, people who want to share information about their drug-taking habits, which makes them more than usually likely to take drugs with their parents. Then, there is the simple fact that they read a magazine about the rave scene, which is notoriously drug-riddled. These aren't typical young drug users; they are enthusiasts, the train-spotters of the drug world.

This statistic is a result of what is known as *sample bias*. The sample was not characteristic of young drug users more generally, and was uncharacteristic in a way that made it more likely to give the result in question.

The need to avoid sample bias when collecting statistics is well-known. The mistake is widespread nevertheless. Newspapers such as the *London Times* should certainly know better, because they frequently publish the results of political polls and sometimes even conduct them. Yet, if it gives a good headline, they are happy to publish the results of a badly biased survey, as the example illustrates.

Our drugs statistic is an example of a common way of ending up with a biased sample, namely, letting the sample choose itself. Those who volunteer to participate in surveys about something are not normal citizens with respect to that something. They are more passionate than most. So, what is true of them is not likely to be true of the wider population.

About ten years ago, the radio and newspapers were thrilled to announce that 40 percent of British women who go on holiday in Spain have sex with someone they had not previously met within five hours of arriving in the country. This statistic was gathered from a survey conducted by a women's magazine. They had invited readers with interesting holiday sex experiences to participate.

More broadly, self-selection bias explains why politicians cheerfully ignore the views of protest marchers, letter-to-the-editor writers, and even party members attending the annual conferences. Only fanatics take part in such political activities, and most voters aren't fanatical.

Most cases of sample bias are quite obvious but some are difficult to detect. For example, you might think it reasonable to take a "snap shot" approach to discovering the average duration of periods of unemployment. Contact some portion of the unemployed population on one day of the year and ask them how long they have been unemployed. Provided the sample is large enough, its average is the average for all those who experience unemployment.

In fact, this sample would dramatically bias the result upward. People who are unemployed for long periods are much more likely to be unemployed on any given day than people who are unemployed for only short periods. Lots of people who have been unemployed for a week are back at work on the day of the poll. So they don't get counted. But everyone who has been unemployed for years is unemployed that day and so they all get counted. To avoid this bias, you need a sample of people, not who are unemployed today, but who have been unemployed at some time in the last, say, ten years. The average term of unemployment in this sample gives a better answer.[2]

Before moving on, I cannot resist mentioning a really egregious case of sample bias. For many years now, it has been taken as a well-established fact that 10 percent of Western men are homosexual. Most believe this statistic but do not know its source. It is Kinsey's *Sexual Behavior in the Human Male*, published in 1948. Alas, 25 percent of the sample used for Kinsey's survey were prison inmates, despite the fact that prison inmates were only 1 percent of the American male population. Since they live in an all-male environment, prison inmates are more likely to have homosexual sex than other men.

It does not follow that less than 10 percent of men are homosexual after all. There were competing forces at work in Kinsey's survey, especially the tendency of people to lie about what was then a taboo activity. So, for all Kinsey's research tells us, we have no idea what percentage of men are homosexual.[3] Looking out the window of my office, I would be inclined to think it is more than 10 percent. Then again, my office is in Covent Garden.

## Anorexia and Other Big Small Numbers

The BMA called for the fashion industry and television to stop focusing on "abnormally thin" celebrities, such

---

2. I owe this example to Steven E. Landsburg, *The Armchair Economist* (New York, The Free Press, 1994), p. 132. I say a better answer rather than the right answer because this sample may not reflect recent changes in average periods of unemployment.

3. Better research by Edward Laumann in 1994 found the percentage of men who are consistently homosexual to be 4 percent.

as Kate Moss, Callista Flockhart, and Victoria Beckham
of the Spice Girls, and for the Government to set targets
on reducing the disease. Anorexia nervosa affects about
2 percent of young women and kills a fifth of sufferers.

—The *London Times* (May 31, 2000)

The British Medical Association (BMA) is always calling on
people to stop doing this or that on account of its dreadful effects
on the health. Normally, their mistake is in thinking that health
is all people care about. I may know that smoking is bad for me
but persist in any case, because I prefer a short and smoky life to
a long fresh one. On this occasion, however, they went wrong on
what should be their home ground, namely, on the medical facts
and figures. The idea that anorexia affects 2 percent of young
women and kills a fifth of sufferers is ridiculous.

There are 3.5 million British women between the ages of fifteen and twenty-five. If 2 percent of them suffer from anorexia
nervosa, that is 70,000. And if a fifth die from it, we should expect
14,000 young women to die from anorexia each year.[4] You will
begin to suspect that something has gone wrong when I tell you
that in 1999 the total number of deaths in women from this age
group, from all causes, including anorexia, was 855. Can anorexia
really kill sixteen times more young women than even die?

---

4. Death rates from a disease are normally expressed annually, i.e., as the percentage of sufferers who will die in a one year period. If the 20 percent death
rate of our example is not annual, but over some longer period, then the number of anorexia-caused deaths each year would be smaller, but not small enough
to save the alleged statistic from massive error. For example, if 20 percent die
in a ten year period, then the number of deaths each year in women from 15 to
35 should be 2,800.

We need not flounder around in the dark. Causes of death are recorded and the figures are available from the National Statistics Office. We can check the number of anorexia-caused deaths in young women. The BMA's figure must be wrong, since no disease can kill more people than die. But how wrong is it?

The figure of 14,000 is more than a thousand times greater than the truth. The number of young women who died from anorexia nervosa in 1999 was 13. Not 13,000. 13.

If I were Callista Flockhart, I'd have sued the BMA and the *Times*. By encouraging the media to stop focusing on her, they attempted to ruin her career, on the bogus allegation that looking at her makes people die of anorexia. Millions of young British women watch Callista Flockhart and at most thirteen die from it each year. That makes Ms. Flockhart safer than crossing the road.

I'm not Callista Flockhart of course, so I did not sue the BMA or the *Times*. But I did write to the editor of the *Times* pointing out their error. Neither my letter nor a correction was published, and I received no explanation of how they could have published such a crazy number. So, I am left to guess at where things went wrong.

My suspicion is that Helen Rumbelow, who wrote the article, suffers from an ailment that afflicts 25 percent of journalists and makes a fifth of them talk nonsense.[5] She has no sense of scale. When numbers get very small or very big, those afflicted lose all sense of whether or not they are reasonable.

We all suffer when the subject matter is unfamiliar. Is forty billion dollars a good price for a space shuttle, or is it a bit over the

---

5. If the BMA isn't too good to make up statistics, nor am I.

top? Most of us wouldn't have a clue. Is 0.01 of a second a reasonable period of time for an electrical impulse to cross a synapse in your brain? Again, unless you are a neuroscientist, you'll have no idea. And anorexia deaths in young women? Well, 2 percent isn't very many. And if only a fifth of them die, that's a very small number: only 0.4 percent. Seems reasonable, doesn't it?

Usually, 0.4 percent is quite a small number. When it comes to deaths in young women, however, it's enormous. Young women hardly ever die. Young men die a bit more. But, more or less, dying is the exclusive preserve of the old. That is something you might have expected the BMA and a medical correspondent from the *Times* to know, and they probably do know it in some general sense. But a very small number like 0.4 percent just didn't ring the alarm bells.

Just as small numbers can be bigger than they look, so big numbers can be smaller. Barclays Bank's profit announcement prompts an outraged newspaper editorial every year. "Three billion pounds profit! And still they shut branches and sack staff. Greedy bastards!" This misses the fact that Barclays is a very big business with many thousands of shareholders. No single greedy bastard gets that £3 billion. In 2002, £3 billion represented a return of only 15 percent on shareholders' investment in the business. That's a reasonable return in these hard times, but hardly scandalous.

The same mistake is at work when you hear all those amazing facts about the cost of repairing the damage done by a hurricane, the economic value of joining the Euro, and so on. A cost or benefit that is spread across many individuals is summed up and presented as a single, shockingly large number. Repairing hurricane damage may cost an amazing $150 million, but it will be borne by ten million Florida taxpayers, costing each of them a much less amazing $15. Joining the Euro really might increase the United Kingdom's GDP by £3 billion per year, as the treasury's recent report claims. But sixty million participate in the UK economy, so each benefits by only £50 per year, or £1 per week.[6]

*

Everyone enjoys being shocked by amazing statistics. But you have to be able to believe them; the fun is wrecked by discovering the statistics are bogus. The brief moment of elation I experienced upon hearing about the promiscuity of English women in Spain was spoiled by discovering the shoddy sample selection that lay behind it. If I hadn't noticed the sample bias, I could have enjoyed the alleged fact for longer. Ignorance is bliss, as they say. But I console myself with the fact that I did not waste the price of an airfare to Spain. Ignorance can also be expensive.

That is the real value of learning to see through bogus statistical claims. You don't make the mistake of acting upon them, by flying to Spain for unlikely sex or pointlessly lowering your prices or supporting silly policies.

---

6. I owe this observation to an editorial by Anatole Kaletsky in the *Times*, June 10, 2003.

# Morality Fever

As a boy, I occasionally told my parents how awful I found some classmate or neighbor. I would list his most appalling characteristics and wait for the parental groans of agreement. But they were never forthcoming. Instead, they always offered some hypothesis as to why the little creep had turned out so (not me, the other kid). His parents had divorced and he felt insecure, his father beat him mercilessly, or something of the sort.

"Maybe," I would protest, "but explaining why he is awful doesn't show that he isn't awful. On the contrary, it assumes he is. So why do you make these remarks as if they count against my point—which was only that he is, in point of fact, awful?" Or words to that effect.

It is bizarre to think that you have refuted a claim by explaining why it is true. How could anyone get so confused as to think this?

Morality fever did it. My parents assumed that I was morally condemning the boy in question. "It isn't his fault" is what they were saying. But I wasn't morally condemning him any more than I would be morally condemning a desert by saying that I find it objectionably dry. The desert can't help it. It is dry nevertheless, and I don't care for it.

Had I told my parents that there is a mountain range in Switzerland, they would not have corrected me by explaining how that mountain range came to be formed. Only in a haze of moral anxiety are people capable of mistaking an explanation for a refutation.

My parents were not alone in suffering from morality fever. It is a widespread malady of the mind, and I suspect it is spreading. An increasing number of opinions and topics seem to raise the moral temperature to a point where the brain overheats.

This chapter is devoted to three more mental malfunctions that commonly occur when morality fever sets in. Being alert to them is important because, where the issues are morally weighty, proper reasoning is required more than ever. Or so I shall argue in the last subsection. Just as all self-help books should begin with a confession, so they should end with preaching.

## What's Wicked Is False

During New Zealand's 1985 public debate on legalizing homosexuality, one of the more peculiar but nonetheless popular arguments was that homosexuality should be illegal because it is unnatural. The argument is peculiar because, whatever is meant by "unnatural," it is silly to think that what is unnatural should be illegal. Miniature golf is an unnatural activity, yet it would be outrageous to criminalize it on that account alone. The same goes for little boys kissing their octogenarian grandmothers, wearing socks with sandals, and open-heart surgery.

Yet, few on the pro-legalization side of the debate pointed this out. Instead they replied that, in fact, homosexuality is natural. This struck me as tactically disastrous, since it tacitly accepted the idea that what is unnatural should be illegal. It is an example of something I have since noted often, namely, a strong bias in favor of arguing about the facts rather than about what follows from them. It is a foolish bias, because it gives the irrational a strong advantage in debates. They need only invalidly draw their favored conclusion from a true premise and an opponent with this bias will be in a hopeless position.

Nevertheless, this is what happened. Most on the pro-legalization side claimed that homosexuality is natural because it has a genetic basis. In 1985, this was a controversial claim and certainly the science involved was beyond the understanding of most in the debate. So it was fortunate perhaps that some had a much simpler approach to establishing that homosexuality is natural. It must be, they argued, since those who wish to keep homosexuality illegal claim it isn't, and keeping homosexuality illegal is obviously wrong.

These thinkers accepted the structure of the anti-legalization argument, but reversed its direction. They agreed that if homosexuality is unnatural it should be illegal. But homosexuality should not be illegal. So homosexuality must be natural.

This approach allows those with moral certainty to discover all sorts of interesting facts about the world without going through the normal rigors of scientific research. By accepting some alleged link between facts about how the world is (e.g., that homosexuality is natural) and facts about how it ought to be (e.g., that homosexuality ought to be legal), those with certainty about the latter are blessed with instant knowledge of the former. Those poor fools struggling in the laboratory to discover a genetic basis for homosexuality; if only they had clear moral vision they could rest easy.

Despite its absurdity, this "moral method" is common where touchy subjects are concerned. The debate about systematic differences in the IQs of different races is the most obvious example. Scientists have published results showing that Asians' average IQ is higher than whites' and whites' higher than blacks'.[1] Most critics of the view reject the finding without any discussion of the research methods or data used to arrive at it. The fact that the finding is agreeable to racists is taken to be a sufficient ground for its rejection.

This reasoning is obviously flawed. How can the fact that racists enjoy hearing something show that it is false? Much of the literature on this topic is at pains to point out that the claim is racist or that those who make it are motivated by racism. But, again, how could this on its own show that whites' average IQ is not really lower than Asians'?

---

1. The IQ controversy was most recently stimulated by R. J. Hernstein and C. Murray's *The Bell Curve: Intelligence and Class Structure in American Life* (New York, Free Press, 1994). Stephen Jay Gould's *The Mismeasure of Man*, (London, Penguin, 1981, Revised in 1996) argues against the "genetic determinism" of *The Bell Curve*.

Besides being logically flawed, the reasoning is also ill-motivated. The opinion in question may indeed be false, but why should an opponent of racism be especially concerned by it? He must accept the racists' reasoning, that such differences would warrant all manner of depredations upon members of races with lower average IQs. That is a dangerous thing to accept, since these differences might turn out to be real.

Those who refuse even to hear an opinion from which others draw unpleasant conclusions tacitly agree that it has the alleged implications. If the opinion turns out to be true, what defense will be left to these good ostriches?

## What's Beneficial Is True

In debates about the existence of God, the religious will often tell you that their faith is a source of comfort in moments of trouble. This is irrelevant. Belief in God would provide this comfort even if God did not exist.

Benefits that a belief would deliver even if it were false are not evidence that it is true. Perhaps everyone who believes she is of above average good looks gains in confidence. This benefit cannot make the belief true of everyone who has it, since this is 90 percent of the population. Believing that he is protected by a guardian angel may make a soldier bolder in battle. But, again, that is no evidence that guardian angels really exist.

This is an obvious point and, outside religion, you would expect the mistake to be extremely rare. But it isn't. Here is Dr. Peter Kenway, the director of the New Policy Institute, defending the U.K. government's discredited disposable income measure of poverty (see pp. 134–137):

It is a simple and reliable statistic, which has played a huge part in propelling poverty high up the political agenda.[2]

Why does Dr. Kenway mention the fact that this statistic has propelled poverty high up the political agenda? How is this relevant in a debate about the measure's accuracy? The statistic has propelled poverty up the political agenda not because it is accurate, but because it is *high*. A more accurate measure that gave lower numbers would have lesser powers of political agenda propulsion. Could Dr. Kenway be making the same mistake as the comforted Christian?

If he accepts that this effect does not show the statistic to be true, then he must think that propelling poverty up the political agenda recommends the statistic even if it is inaccurate. That is a strange idea. If poverty is not really as big a problem as the statistic indicates then, surely, it should not be so high up the political agenda. The proper policy priorities depend on the facts. The idea that we should begin with the priorities and then tailor our view of the facts to suit them is absurd.

Clare Short, the former International Development Secretary, accused the Prime Minister of doing just this when she described his claims about the military threat posed by Iraq as "an honorable deception." He started with his policy of invading Iraq and then tailored his (and our) view of the facts to suit. Who knows if she was right. But, if she was, "honorable" is a peculiar description of the behavior. "Fevered" might have been better.

---

2. The *Guardian*, April 19, 2002.

We all sometimes indulge in such shenanigans when defending a position we favor. But we know we shouldn't, even when we think our position is righteous. If defending it requires misrepresenting the facts, then it cannot be correct after all. Righteousness isn't above the truth.

## The Meek Shall Inherit the Truth

In 1985, the New Zealand Rugby Union planned a tour of South Africa. Anti-apartheid organizations in New Zealand strongly opposed the plan. As part of his campaign to rally opposition to the tour, John Minto, the leader of HART (Halt All Racist Tours) spoke at Auckland University.

After his speech, a student took issue with some of the stronger measures recommended for dissuading the national rugby team from going ahead with the tour. The civil liberties of rugby players was the topic of discussion, but the details do not matter. The interesting point in the debate came when Mr. Minto, having become impatient, made what he took to be the killer point. The blacks in South Africa were starving and did not want this tour to go ahead. He listened to them, and not to white, middle-class suburbanites.

The WMCS student replied that the poverty of South African blacks hardly made them more likely to be right on the points of political philosophy under discussion. The cacophony of booing and jeering elicited by this remark ended the debate in Mr. Minto's favor.

But the student was right. However unjust apartheid, it didn't bestow on its victims the power of infallible truth. Moral outrage at someone's mistreatment does not oblige you to agree with everything he says. Mr. Minto had morality fever.

He was not unusual in exhibiting this symptom of it. In the academic humanities—anthropology, sociology, literary criticism, and the like—it is almost an orthodoxy that the opinions of those who have suffered from Western imperialism are immune from criticism. Say that the traditional beliefs of some such society are false, and you will soon be accused of intellectual or cultural imperialism.

Whether or not it is a kind of imperialism to criticize the views of other cultures is a matter beyond the scope of this book. The important point here is only that this immunity from criticism cannot have the source many of its advocates allege it to: namely, that truth is culturally relative.

Cultural relativism about truth, which is popular in the humanities, is the view that any belief widely held in a culture is automatically true in that culture. In Iran, it is generally agreed that there is only one god. So, in Iran, it is true that there is only one god. In Papua New Guinea, the people tend to believe in many gods. So, in Papua New Guinea, there are many gods. Every culture has the power to make its own truth. Disagree with the consensus and you must be wrong, by definition of truth.

Cultural relativism is so absurd that it is hard to believe anyone can be so fevered as to assert it. If it were true, gods, planets, bacteria, and everything else would come into and go out of existence according to what people generally believe to exist. Which they obviously do not. And statements that contradict each other, such as "There is only one god" and "There is more than one god," would both be true, since the beliefs common in different cultures are often contradictory. But, it is impossible for contradictory statements both to be true. Don't be fooled by the "in Iran," "in Papua New Guinea," "in nineteenth-century France," and similar tags relativists place at the end of claims about what is true. Iranians believe there is only one god, not only in Iran, but everywhere. This belief cannot be true in Iran but false in Papua New Guinea. If it is false anywhere it is false everywhere.

Many enjoy feeling guilty about misdeeds they didn't do, such as colonizing Africa or denying women the vote. I have even seen undergraduates, who I was fairly certain were virgins, marching with placards declaring "I am a rapist." Who would deny people such innocent pleasures? But, like all pleasures, *faux* guilt must be enjoyed responsibly. You mustn't allow yourself to get so carried away with it that you contract the fever and start believing that your victims are infallible.

## Be Serious

The fallacies discussed in this book have several reliable sources: Congress, talk radio, and newspaper editorials will give you all the material you need to hone your skill at spotting them. But, if you haven't much time and seek a really condensed dose of muddle, I recommend you write to the BBC and ask them for recordings of their Radio Four panel discussion, "The Moral Maze."

The panelists on this program are intelligent, well-educated people—mainly academics, religious professionals, and politicians. They are quite capable of reasonable thought. Put a few of the day's weightier moral topics in front of them, however, and they soon show all the signs of the fever. If you can bear to listen for long, it becomes clear that many are more interested in displaying their concern and sincerity than in arguing cogently. Indeed, they seem to believe that genuine concern licenses irrationality. "You can't argue with his sincerity" is the reaction they seek. And in seeking this they resemble many of their listeners.

The idea that you can't argue with the morally sincere, that caring licenses irrationality, is as pernicious as it is popular. It displays a lack of moral seriousness. If the matter at hand is something you genuinely care about, then you should seek more than ever to believe the truth about it. And rationality is merely that way of thinking that gives your beliefs the greatest chance of being true. To dispense with it on the ground that you care is preposterous. As the moral temperature rises, so should our devotion to the truth and hence to proper reasoning.

The idea that sincerity may substitute for reason is founded on an egocentric attitude toward belief: that what I believe is all about me, not about reality. What matters is not that the position I favor will have the best or the intended effects, or that the problems I worry about are real or grave, but only that I hold my position from the right sentiments, that *I* am good.

A similar egocentric disdain for the truth underlies many of the more obvious fallacies discussed in the early chapters of this book. People will hold an opinion because they want to keep the company of others who share the opinion, or because they think it is the respectable opinion, or because they have publicly expressed the opinion in the past and would be embarrassed by a "U-turn," or because the world would suit them better if the opinion were true, or . . .

Perhaps it is better to get on with your family and friends, to avoid embarrassment, or to comfort yourself with fantasies than to believe the truth. But those who approach matters in this way should give up any pretensions to intellectual seriousness. They are not genuinely interested in reality.

Nor are they genuinely concerned about the welfare of others. For we all live in reality, even if we might wish it otherwise. To know what is in the best interests of those you care about you need to understand the world in which they live. If heaven does not really exist, for example, then those deprivations the religious recommend as the path to it are not really in their children's best interests. If they are seriously concerned about their children, they should be serious about the existence of heaven. And if this is true for religion, it is even more obviously true for physics, biology, economics, psychology, medicine, and everything else on which people have opinions.

Separating intellectual from moral seriousness is harder than those who are intellectually frivolous may care to admit.

Jamie Whyte is a former lecturer of philosophy at Cambridge University and winner of *Analysis* journal's prestigious prize for the best article by a philosopher under thirty. He has published numerous articles—mainly on the subject of truth—in journals such as *Analysis* and the *British Journal for the Philosophy of Science*. He is from New Zealand and now lives and works in London.